Global Environmental Issues

Empirical Evidences

First Published — [illegible]

ISBN: 978-81-8387-369-7

Published by

SERIALS PUBLICATIONS
4830/24, Prahlad Street, Ansari Road
Darya Ganj, New Delhi-110002 (India)
Phone: 23245225 Fax: 91-11-23272135
E-mail: serials@satyam.net.in

Global Environmental Issues

Empirical Evidences

Editor

B. Suresh Lal

SERIALS PUBLICATIONS
NEW DELHI (INDIA)

First Published - 2010

ISBN: 978-81-8387-369-7

Published by

SERIALS PUBLICATIONS
4830/24, Prahlad Street, Ansari Road
Darya Ganj, New Delhi-110002 (India)
Phone : 23245225 Fax : 91-11-23272135
E-mail: serials@satyam.net.in

CONTENTS

FOREWORD

Our mother earth is the most precious gift of the Universe. It is sustenance of nature that is a key to the development of the future of mankind. It is a befitting topic to address the present global environmental scenario. The earth is the only planet in the solar system known to have liquid water on the surface and to be geological active. It is also the only place in the Universe known to support the life. The degradation of our environment is linked with the development process and the ignorance of people about retaining the ecological balance. To reduce the negative impact on land and by the in surrounding atmosphere we need to do a lot. The balance of the eco-system is out of gear now. This problem questions the very survival of the human and all other living organisms. The life supporting environment is also heavily disturbed. To set it right, efforts are being taken at the global level to solve the global environmental problems.

The present edited volume of Dr. B. Suresh Lal on "Global Environmental Issues: Empirical Evidences" is useful in understanding the various dynamics of environmental issues in the contemporary world society.

Dr. Lal, worked for this book, all the papers contained in this volume, will be a value addition to the growing literature on this subject, by a variety of scholars on very wide ranging issues. That there are twenty papers is both a challenging task but also a limitation as it is difficult to put all of them in a running theme. The papers contributed in this volume are written by reputed scholars from various universities and research institutes and from various countries. The paper writers' focused on various global environmental issues in the contemporary society of the world. Almost all papers are based on empirical evidence.

The issues covered relate to global environment. This includes Environmental degradation, impact of environmental pollution, impact of aquaculture, ground water resource mapping, land coverage mapping, conservation, women and natural resources energy, entrepreneurship, forest development, agricultural development, rural development, environment management accounting and environmental protection. In nutshell all the papers compiled in this great little book discuss the environmental issues and these papers put together are multi-dimensional and raise certain questions about core-issues of our environment. The content and concerns are very relevant. That such a range of issues could be brought under a volume i.e., Global Environmental Issues: Empirical Evidences by Dr. Suresh Lal, a young scholar is an accomplishment by itself.

I wish to congratulate Dr. Lal, who has done a commendable job by putting together into a cohesive volume twenty different studies in a common conceptual framework and drawing lessons from them. I am sure, the book will be of immense value to all those interested in environmental change and degradation, policy makers and future researchers.

Prof. N. Linga Murthy
Vice-Chancellor
Kakatiya University,
Warangal, AP, India.

ACKNOWLEDGEMENT

In the preparation of this book I have been benefited from a large number of sources. First of all, I acknowledge with deep sense of gratitude my warmest appreciation to the Vice-Chancellor, Kakatiya University, Warangal Prof. N. Linga Murthy for giving vent to his impression about the present work in the foreword of the book. I greatly cherish his kind gesture of encouragement and feel extremely beheld to me.

I am also immensely grateful to Prof. G. Rajaiah, Dr. M. Ravinder, Dr. S. Jyothi, Dr. AV. Ramana, Mr. Suresh Modem and all those who have been consistently encouraging me right from the start of the work.

I sincerely express my heart felt gratitude to Dr. Ramkishan Ajmeera, Mr. Chandra Mohan, Mrs.Padma Banoth, Pragna Rathore, and Sushanth Chaitanya Rathore, who has encouraging and inspiring me in academic work.

I sincerely thank to the paper contributors for their qualitative evaluation in the form of studies and insightful analysis.

I acknowledge with deep sense of gratitude to M/s Serials Publisher of Mr. SK Jha for giving me editing opportunity to this volume and who has very kindly agreed to publish this book.

Dr. B. SURESH LAL

LIST OF CONTRIBUTORS

Akongbowa B. Amadasun, Ph.D (AAU), Benson Idahosa University, Benin City, Nigeria.

Alok Prasad Das, Department of Chemical Engineering, National Institute of Technology, Rourkela-769008, India.

A.S. Chandra Bose, Lecturer, Dept. of Civil Engineering, Govt. Polytechnic, Warangal, India.

Emmanuel O. Eyo, Department of Agricultural Economics and Extension, University of Calabar, Calabar, Nigeria.

Esam Ismaeel Jassim, Research Scholar, Department of Chemical Engineering, University of North Dakota, Grand Forks ND 58202-7101.

G.D. Kharat, Lecturer, Dept. of Economics, Arts & Commerce College, Yeola, Nashik, India.

G.K. Viswanadh, Professor and Head, Dept. of Civil Engineering, J.N.T.U. C.E., Hyderabad, India.

G. Rajani, Centre for Environment, Institute of Science & Technology, Jawaharlal Nehru Technological University, Hyderabad-500072, A.P, India.

I. Sundar, Reader in Economics Directorate, of Distance Education Annamalai University.

K. Kiran, Centre for Environment, Institute of Science & Technology, Jawaharlal Nehru Technological University, Hyderabad-500072, A.P, India.

Lal B. Suresh, Associate Professor, Department of Economics, University Arts & Science College, Kakatiya University, Warangal, AP, India.

M. Anji Reddy, Centre for Environment, Institute of Science & Technology, Jawaharlal Nehru Technological University, Hyderabad-500072, A.P, India.

M. Ganesan, Lecturer, Department of Economics, Aditanar College of Arts and Science, Tiruchendur, Tamilnadu, India.

M. Sumathy, Lecturer in Commerce, Periyar University, Salem-636 011.

N. Ahalya, Energy & Wetlands Research Group, Centre for Ecological Sciences, Indian Institute of Science, Bangalore 560 012, India.

Padma Seragadam, Department of Chemical Engineering, National Institute of Technology, Rourkela-769008, India.

Pradip Swarnakar, Centre for Interdisciplinary Studies in Environment and Development, ISEC Campus, Bangalore.

Prafull K. Singh, Center of Remote Sensing & GIS, School of Studies in Earth Science, Jiwaji University, Gwalior (M.P), India.

Rameshwari Pandya, Senor Lecturer, Department of Extension and Communication, Faculty of Family and Community Sciences, The M. S. University Baroda, Gujarat.

Ravindar. M, Lecturer, Department of Economics, Govt. Junior College, Parkal, Warangal, AP, India

R.B. Dhande, Reader, Dept. of Commerce, S. P. H. Mahila College, Malegaon Camp, Nashik, India.

R.D. Kanamadi, Department of Zoology, Karnataka University, Dharwad, India.

R. Mohanraj, Research Scholar in Economics, Annamalai University.

Shouju LI, Yingxi LIU, He YU, State key Laboratory of Structural Analysis for Industry Equipment, Dalian University of Technology, 116024, China

Siddhartha Sarkar, Faculty, Department of Commerce, Dinhata College, Dinhata-736135, Cooch Behar, West Bengal.

S.S. Asadi, Scientist, Centre for Environment, Institute of Science and Technology, Jawaharlal Nehru Technological University, Kukatpally, Hyderabad-500085, India.

S. Srinivasulu, Associate Professor, Dept. of civil Engineering, JNTU, College of Engineering, Kukatpally, Hyderabad-500085, India.

Susmita Mishra, Department of Chemical Engineering, National Institute of Technology, Rourkela-769008, India.

T. Sezhiyan, Lecturer in Economics, Directorate of Distance Education Annamalai University.

T.V. Ramachandra, Energy & Wetlands Research Group, Centre for Ecological Sciences, Indian Institute of Science, Bangalore 560 012, India.

U.C. Singh, Center of Remote Sensing & GIS, School of Studies in Earth Science, Jiwaji University, Gwalior (M.P), India.

EDITOR'S NOTE/INTRODUCTION

From Kyoto Protocol to Copenhagen, the World Environmental Summit of UN is going to be taking place within a week at Copenhagen, Denmark on 7th December-2009, at the same time I have started final editing of the book "Global Environmental Issues". As we are aware of the problem of green house gases releasing into atmosphere by many advanced countries, they have to take more responsibility to this adverse affect on the earth. A healthy environment and a healthy economy is the slogan of the Nations. With this background the book "Global Environmental Issues: Empirical Evidences" is coming out to create awareness among the students, scholars, teachers, policy makers and activists of the environment. This volume is mainly based on empirical studies conducted by researchers, who have zeal to present the issues from various dimensions and a new angle.

The Nigerian forest which provides base for economic sustenance, scientific research, tourism and providing commercially important products is loosing its resources and degrading rapidly posing serious threat to life and environmental protection and sustainable forest development in Nigeria in the 21st Century. The author *Akongbowa B. Amadasun* presented underlying causes for deforestation which examining policy areas of government forest, management development and attempted to analyze emerging issues in the 21st century, suggesting measures for sustainable forest management and environmental protection. The author started a debate for an alternative order in the place of existing ones.

Emmanuel O. Eyo studied agriculture sector growth in Nigeria, relating to macro economic environment. The historical evolution of Agricultural growth has been presented in the study. The author assessed the effects of Macro economic policies on the agriculture sector by using regression equation. The study concluded that the Structural Adjustment Programme (SAP) and changes in macro economic policies changed the macro economic environment favourable to the growth and development of the agriculture sector in Nigeria.

In the paper titled impact of WTO on agricultural marketing, the authors *R.B. Dhande* and *Kharat* suggested measures to improve agricultural marketing system keeping in view the status of Indian farmers and agricultural sector in the light of Agreement on Agriculture (AoA) of WTO which is biased in favour of developed countries.

In environment management accounting (EMA) an overview, *M.Sumathy* analyzed various aspects involved in environment management accounting (EMA). In the wake of growing environmental awareness, it was emphasized the importance of EMA for companies to reach goals of saving money, environmental costs and reducing environmental impacts. It also presented the practice of EMA in some companies in Japan and USA.

Rameshwari Pandya in environment Vs development, can we have equilibrium? focused on environmental crisis which is the result of excessive appropriation of natural resources in pursuing growth oriented development. The author also discussed environmental degradation reaching the level of crisis in developing countries. Pointing lapses in acquisition of farm lands for SEZs and implementing environmental protection acts effectively, the author suggested ways and means that can achieve equilibrium between environment and economic development where development does not degrade environment and environment does not refund development.

In the name of protecting endangered species as part of environmental protecting measures, statutes made and ways of implementing these statutes people of some sections of the society are often jeopardized. The laws made for protecting olive ridley turtle along the Orissa coast and way of implementing these acts are negatively affecting fishing community people, specially, small fishermen. Various agencies working for protection of turtles are also silent about the grievances of fishermen. The double standard view of government that encourages industrial groups to carry out activities along the coast causing damage to marine environment and species and restricts the fishermen to livelihood is also negatively affecting the fishermen. In the light of the grim reality, a more comprehensive policy involving all stake holders should be taken to make the idea of "live and let live" meaningful for both humans and turtle. This is the analysis of *Pradip Swarnakar* in the who needs conservation and from whom? olive ridley turtle conservation in Orissa sea coast.

In the study on environmental degrading in Barhur Sipcot Industrial Region, *I.Sundar* and *R. Mohanraj* analyzed positive and negative externalities of sipcot establish Burger industrial region in Tamilnadu by using primary data. The negative externalities in the form of land, water, air pollution and consequent problems causing increased health expenditure, loss of properties and traditional employment base, decrease in crop yield etc upset positive externalities whatever little created by industrial region. The study concluded that the government and industrial management should initiate concrete measures to provide sustainable livelihood to people victimized by the industrial complex and ensure people lead peaceful life which they enjoyed earlier.

Lal B.Suresh and *Ravindar* analyzed, environmental pollution impact on health and economy an overview study". They focus how pollution is creating the problem of environment as well as health issues. Environmental pollution affects not only human health but also biodiversity and economy. They also analyzed contamination of water, air and land create many communicable diseases, thereby cause monetary loss.

Esam Ismaeel Jassim in the article modification of particle deposition model in turbulent pipe flow, described the particle migration and deposition keeping in view its significance as heat exchangers, health effects and cause environmental risks.

Susmita Mishra and others indicated in their study isolation and characterization of nickel-resistant bacteria from electroplating waste, that the electroplating waste containing nickel along with other heavy metals can be removed at its lower concentration effectively using the identified species staphylococcus aurous.

In the acquifer parameter identifiable with simulated annealing algorithm and its application, the author *Shouju* LI and others have presented a method based on simulated annealing algorithm for obtaining the parameters of linear ground water flow model and dissolved contaminant plume useful to find solutions for ground water seepage problems.

N.Ahalya and others explained in their article biosorption of chromium (VI) by *tamarindus indica* pod shells that tamarind pod shells have been used for the absorption of Cr (VI) and its utility as biosorbent will be economical and be viewed as part of waste management strategy. And also they opined that tamarind pod shells are advantageous over the protein rich algal and fungal biomass.

S.S.Asadi and others in their study titled aquaculture impacts on ground water quality in Pennar river basin using remote sensing and GIG: a model established that the groundwater quality in the study area is adversely affected in terms of increased TDS, hardness and others, by using data generated remote sensing and geo information system and suggested coordinated action by geoscientists, engineers, environmentalists, planners/ policy makers to improve the conditions in the study area.

A.S. Chandra Bose and *G.K. Viswanadh* in their research Analysis of metrological drought using GIS assessed meteorological drought and prepared thematic maps using Arc GIS 9.2 in Godavari basin in the Deccan plateau covering large areas in the states of Maharashtra, Madhya Pradesh, Chattisgarh, Orissa, Karnataka and Andhra Pradesh. They concluded that GIS could be effectively used for meteorological drought assessment with greater efficiency and accuracy.

By using data from survey of India and satellite imaginaries from National Remote Sensing Agency and other agencies *S.S.Asadi* and *S.Srinivasulu* prepared ground water prospect maps and identified sites for construction of recharge structures to increase ground water levels in the part of Hyderabad and in Mahabubnagar district of Andhra Pradesh. In their study titled as integrated approach for ground water resources mapping using remote sensing and geographical information system.

Siddhartha Sarkar in his study, women, natural resources, environment and livelihoods diversity analyzed environmental degradation keeping in view the reliance of poor on natural resources for their livelihood. The study revealed that women and children were badly affected by environment degradation. The study suggested strong political will and involvement of rural poor and women in social decision making to protect environment for sustainable economic development in general and for sustainable livelihood of rural poor in particular.

Keeping in view the importance of energy sector in energy based present day economies and energy needs of women in performing their household duties and to carryout other enterprising activities and key role that women can play in energy sector, the author *Siddhartha Sarkar* in women as energy entrepreneurs, emphasized the need to provide modern forms of energy to women and involve women in every sector for effective management in the light of Millennium Development Goals (MDGs) adopted by the UN General Assembly in 2000. The author analyzed women in relation to energy in the Asian and African countries.

Prafull K. Singh and *U.C. Singh* in their work land use/ land cover mapping using digital image processing of IRS-ID satellite data a case study of Morar watershed, district Gwalior, demonstrated the application potential of remote sensing techniques in assessing land use of the area. The agricultural area in northern portion and forest area in southern portion of the watershed affected by severe soil erosion due to the occurrence of shales and having steep slope. They suggested for urgent need to manage soil erosion applying various techniques of soil conservation and large-scale afforestation.

M.Ganesan in the study trends in agriculture during pre and post reforms period in India estimated the food productivity, per capita availability of food grains and area under food crops during pre and post reform period in Indian agriculture sector and pointed out the negative impact of liberalization, privatization and globalization policies on Indian agriculture. The study suggested increase in public investment to develop rural infrastructure to benefit small and marginal farmers.

T.Sezhiyan has analyzed the regional variation in rural development in Cuddalore district on different chosen socio-economic indicators like density of population, sex ratio, average size of household, percentage of workers to total population etc. Inter-block variation and inter-village variation have been observed with respect to rural infrastructure development in the district.

It is in this background that the present volume attempts to analyze and uncover various issues relating to environment such as the process, content and other related dimensions. The contributors being from diverse disciplines, it is hoped that the volume contributes to the scantly explored area of study of global environmental issues in India and among the world countries.

1

ENVIRONMENTAL PROTECTION AND SUSTAINABLE FOREST DEVELOPMENT IN NIGERIA IN THE 21ST CENTURY

Akongbowa B. Amadasun

INTRODUCTION

Post independence management of the Nigerian environment indicates massive exploration of the environment (natural resource inclusive) for biological survival; economic transaction and international trade; Government environmental policy mistakes and inconsistencies/distortion and sustainable development deficits.

There are many factors that are responsible for environmental degradation. However, since this paper is essentially for forest exploitation and conservation, the discussion that follows will be on the Nigerian forest and its natural resources.

Nigeria has a total landmass of about 960,000 km of which 40% is classified as forestland with high forest zone of about 37% and savannah woodland of about 67%. Annual wood consumption in Nigeria is put at 5.1 and 88 million cubic meters of which about 80% is consumed in the rural household for cooking and building and construction purposes, resulting in acute ecological problems of forestation and erosion.

As at today Nigeria forest provide a base for economic sustenance, scientific research, tourism, and providing commercially important products such as logs and timber, nuts, honey, palm hearts, rubber resins, species of wildlife, insects, birds, etc., yet forest issues are undervalued in terms of national policies and priorities.

Statistical data indicates that desertification has already affected 350,000 sq.km advancing at 0.6km per annum, with the depletion of the forest at 3.5% annually, 3.7 million hectres of forest are loss between 1937 and 1984, undisturbed forest decreased at 2.9% in 1976/78 to 1.9% in 1995 (Akinsani F.A)[2].

As at today only 4% of rainforest is left, and of the 11,000 species of animals and plant threatened worldwide 484 species come from Nigeria. While these figures may be disputable one fact that is clear is that Nigeria forest and its resources are shrinking at an alarming rate, posing grave danger to life and environment. The emerging questions therefore are:

- What are the policy areas of government forest management development?
- What are the real underling causes of rapid deforestation?
- What are the strategic challenges of sustainable forest management in Nigeria?

Policy Areas

In sum, policies of forest management/development Nigeria center around:

I. Chasing illegal log fellers.

II. Creating forest reserves that were seldom protected.

III. Formulating policies around revenue mobilization from the forest with little or no plough back forestation/afforestation.

IV. Soliciting for international funding that may not materialize misappropriated or diverted to non-forestry sector funding.

V. Running a grossly dysfunctional and illequiped forestry ministries/department and for weak institutional/regulatory framework.

VI. Signing at some international forestry/environmental conventions.

Real and Underlying Causes

Awake[3] states that the real and underlying causes of forest degradation/destruction as:

- Government pillage of forest in quest for power and riches.
- Official corruption which has written death warrants for forest, logging permit is worth big money that for a token dishonest officials have been known to give short term concessions to companies to plunder the timber without regard for conservation/regeneration.
- Shifting agriculture and changing woodlands threat to forestry or incompetent officials signing away forests that can never be recovered.
- Introduction of chainsaws and bulldozers into clear-cut of forestry after the second World War, which assisted/aggravated the speed and volume of deforestation.
- Waves of urbanization process that increased demands on forest products for building, construction and recreation.
- Civil and armed conflict, which exacerbated the problem of illegal logging and poaching of endangered species for ivory and meat.
- Globalization, which has resulted in the industrialized/rich west exploiting developing countries (Nigeria inclusive) fragile forest/environment.
- Petroleum exploration, mining and/or government politically anchored and haphazardly organized large-

scale agricultural projects, which have largely aggravated forestation/degradation.

- Commercial timber exploration, oil plantations, and life stock production, which is responsible for the continuous high rate of deforestation.

Strategic Challenges

1. A Sustainable Forestry Management strategy with the capability to support life, and a more environmental and socially equitable approach to forest management that is not driven by sheer providence and greed but by concrete actions with resolve at stakeholders to act locally while thinking globally, and to recognize the nexus between sustainable economic development and forest/environment protection.
2. To adopt policies and actions informed by sincerity towards the cessation and possible reversal of the sleeping destruction and degradation of the forest/environment of these states and the review of forestry and environmental laws to reflect the urgency, legal and institutional framework for effective management and protection of natural resources (and forest and waterways).
3. Actions that should address proposals for change in the context of strategic policy initiatives aiming to strategically and optimally re-align Nigeria's forest resources and environmental integrity, attain sustainable development and equip the sector for integrative growth challenges.

EMERGING ISSUES

Based on the implication of the Nigerian deteriorating environmental cum-forest health and safety, the emerging issue is what are the integrative policy issues that its potend the Nigeria in the 21st Century? The paper position here is that this could be best answered/presented through an array of questions, amongst others. Viz:

- What type of forestry cum environmental policy objectives should Nigeria pursue? Should it be short, medium or long term policy objectives or goals?
- Is Nigeria currently having in place forestry cum environmental policy?
- What are the institutional/regulatory frameworks in place for providing for and/or enforcing sustainable energy, construction and urbanization policy induced forestry cum environmental health and safety? Are they appropriate and adequate? And how efficient/effective?
- What is the level of intellectual contribution/participation of the public on issues of sustainable forestry/environmental health and safety matters?
- Are there in operation policies that foster excessive resources use?
- To what extent do environmental taxes on activities of loggers and other users of our forest and environment foster/induce environmental health and safety hazards?
- Do environmental interventionist agencies have the institutional capacity – is there clarity of mandate, adequate resources, technical capacity with authority to operate across disciplinary and jurisdictional lines and accountability for results?
- What should be the incentive policies of a sustainable forestry public sector, the private sector and the affected communities/inhabitants collaborative and/or co-operative resolve efforts?
- What patterns of interactions are manifest and among levels of forestry/environmental sector policy actors? What is the nature of these interactions? How are they managed for effective and efficient realization of environmental health and safety?
- What strategies/policies and accompanying investment should be legislated for in order to

prevent/mitigate/eradicate mindless profit and urbanization induced forestry/environmental health and safety problems? Are resources adequate to sustain the proposed strategy and technological investment?

- Is the public sector the best place to manage sustainable forestry/environmental health and safety problems given the capacity and capability constraint or inadequacies?
- What methods should be in place for monitoring/evaluating the failure and success of a sustainable forestry/environmental strategies for preventing environmental quality determination?
- Who should be held accountable for operators failure to adhere to an environmental assessment impact? Should it be the firms or the regulatory interventionist agencies or both?
- What should be the role of professional bodies or associations and ethics in initiating and implementing environmental friendly health and safety sustainable forestry policy induced programmes?
- What should be the tools and methods of compensation and regeneration for forestry policy induced environmental health and safety.

Questioning the status quo and being rigorous in analysis will ensure more thoughtful and innovative approach to building a sustainable forestry environmental quality management.

SUSTAINABLE FOREST/ENVIRONMENT QUALITY MANAGEMENT

Many approaches to the concept of sustainability are bound, based on proper management of forestry/environmental resources in a manner that meets intra-and inter-generational equity needs. This requiring the balance of human needs against the potential that the environment has for meeting them. Consequently, the implication this holds for a sustainable forestry/environmental policy is that the policy should identify the correlation between health and safety requirements of Nigeria's environmental (forests inclusive) and the urgent transition of energy activities to support sustainable development and/or economic growth.

Many definition of sustainability are anchored on development. A few of them include that of world commission of environment (1987), note by Uniamikogbo Sam etal sees sustainability as meeting the needs of the present without compromising the ability of future generations.

Akinyide Osuntokun (1997) sees sustainable development as that development that meets the present population without compromising the ability of future generations to meet their own needs. This the emphasis of the above definition is the need of existing and future generations. Oniamikogbo (1999) citing Morvaridi (1996) and egbon (1996) indicates that sustainability should incorporate a multi-disciplinary approach which embed the under listed:

(i) Concern for economic and social viability of future generation; that is, intergenerational equity;

(ii) Alleviation of poverty and concern for social life.

(iii) Local participation in planning and execution of development projects; and

(iv) Positive impact on environment and resource use.

He further assets that in this context, the concept of sustainable development gives a wider dimension to the problems of development that of traditional strategies based on development economics. Besides, the quantitative dimension of sustainability involves assessment of society's long-term ecological, social and actual potential for supporting economic activity and structural change. That the use of policies can contribute to a new and awesome load of environment protection in an efficient allocation of scarce resources. Thus, this calls for adequate social and environmental cost- benefits analysis that will require users to meet the full environmental and social costs of resources used in addition to the cost of extraction.

As forestry/ environmental per capacity consumption increase and export that are dependent on our environment grows the style of management of the environment which will determine the sustainability of the environment. However, where exploitees are not control as in the case of forest, their activities can lead to environmental and/or forest degradation and bankruptcy. To meet the challenges of ameliorating the degradation resources needs to rationally and functionally utilize within the framework of sustainable development.

Egbon C. peter contributing to environmental integrity (ensure policy/sustainability incentives) Pins that ecological objectives. Remains on of the three trade offs of sustainable development (forest inclusive). Others he stated to include economic objectives and social objectives.

Ecological objectives study reveals that it has now become the major objective of development. By 1980 body evidence has accumulated. With environmental depreciation to which exploitation is a contributor as a major barrier to development. Consequently, the view is that sustainability should therefore promote continuing attempts to integrate environmental concerns into conventional economic decision making mainly by valuing environmental assets and inputs of developments effects and using them in procedures ranging from cost benefit analysis to environmentally adjusted national accounts at macro economic level. However, there is need to note that some functions of the environment cannot be valued in monetary terms, other techniques like multilateral analysis will need to be use to trade off non-comparable objectives taking a total implications of environmental protection and sustainable forestry development, the focus of the biological and physical systems is particularly important due to viability of the overall system. In this view protecting the environment is a key object. This emphasizes protecting the resilience and dynamic ability of such systems to adapt to change rather than conservation of an ideal static state.

Taking the above into account, sustainable environment (forestry inclusive) quality management. Reflects that:

1. Sustainable development demands compensation for the opportunities forgone by future generations because today's economic activities (forest exploration inclusive) changes the level of comparism of biodiversity in a way that will affect the flow of vital future ecological services, and now the options available to unborn generations. This holds even if positive rates of economic growth indicate an increase in the instrumental (or use value of options currently available of options currently available.
2. Reconciling these various concepts and operationalising them as a mass to achieve sustainable development is a formidable task. The diversity of short terms needs and concerns as well as long term goals.
3. Lastly the identification of sustainable options, require.
   - Good understanding of the physical, biological and social impacts.
   - Better estimate of values of damage to the environment that help to improve the design of policies and projects and environmentally sound investment decision(s).
   - Development of policy tools and strengthening of human resources and institutions to implement viable strategies and manage resource on a sustainable basis.

In the past three and half decades in Nigeria concern on environment are products of the evidence that natural processes are being distorted by productive activities to such an extent that the quality of life or even life itself is threatened. Therefore environmental quality depends on the balance between economic development and preserved quality availability in any given locality. Consequently Nigeria forest and environment sustainability in must be seen in terms of the need for future generations to meet their own needs. Thus there must be a balance between economic growth and sustainable development. As maintenance of capital of which are made up of human capital, human made capital and natural capital (say natural resources-forestry inclusive).

For sustainability to be said to be operational in the forestry sector the criteria a which it must fulfill are:

- Concern for economic and social viability of future generation.
- Inter-generational equity
- Alleviation of poverty, concern for social welfare.
- Local participation in planning executing of development projects (i.e., let people be part of the issues under consideration, allow them to contribute so that there will be sense of belonging, and this makes for sustainability.
- Positive impart on the environment and resources use.
- In his contribution all the issue Egbon (2003) stated that environmental quality policy planning stated that:
    1. It is expected to induce macro- economic stability through demand management in the short terms as a basis of economic growth and balance of payment equilibrium in the medium term.
    2. There is equally enough evidence that macroeconomic and sectorial policies within the framework of structural Adjustment programme (SAP) significantly impact on resources use. An approach of sustainable development will therefore make imperative the integration of growth structural change with environmental protection and improvement.
    3. Different policy effect also manifest in different form of environmental resources integration.

Sustainability demands that government takes care of its citizens. This certainly is only if there is output growth. Where there is no output growth, obligations to citizens may not be implemented. Besides, when economy grows its output grows. But resources may be used up as economy grows e.g. forest, water, and for instance dry wood was used as poles so far because it does conduct electricity but now such poles are very tiny. Now the forests for such trees are so demanded that they cannot produce a good desirable size of poles. Consequently, the talk of growth is a talk for down grading natural resources. The sum is that the more people are poor the greater the tendency for the forest and environment (natural resource) to get degraded. Because the tendency is for poverty to tom on the forest and for environment for survival. Where poverty or endemic forest exploration should not be left to individual discretion. Largely, because resources will be sub-optimally allocated where individual discretion is allowed.

RECOMMENDATION AND CONCULSION IN ENVIRONMENTAL PROTECTION

The task ahead in sustainable forest development (SFM) is superlatively great not daunting. The strong dependency on the forest for sustainable livelihood, and alarming forest destruction and environmental degradations signals an impending ecological and human disaster. Calling for a dynamic policy prescription/initiative; genuine global partnership in market driven environment; a long overdue forestry and environment focused long range vision for sustainable integrative intersectoral development; and the need to abandon short terms, that is money subsisting and progress by crisis management for a long range economic forestation/reforestation plan and agricultural technical/ industrial nexus as a basis of national rendezvous.

Apparently this requires urgent search for effective action requiring sustainable and collaborative efforts of all stakeholders, and emerging analysis of where do we go from here. Should it be a case of new architecture that significantly reflects sustainable development and poverty alleviation needs of Nigeria or do we build on some of those things that are already in place and discard those things that have not worked well? In the light of the above, the paper adopting isimekha Z. A. classification, proposes the foiling as part of an inexhaustible policy recommendations for the way forward.

General

- Allocation of sites of forestry should used plan derived from proper land use classification.
- There should be effective mechanism for capacity building, institutional strengthening and farmer training to facilitate realization of forestry projects;
- Forestry projects should be adequately and timely funded;
- Special forest areas such as strict nature reserve, man and biosphere reserve should be set aside as inviolate plots for research on the natural ecosystem and conservation of biodiversity.
- Continuous education, information sharing and public awareness campaigns on conservation and utilization of forest resources should be intensified.
- Forestry should properly integrate with food, fish and livestock production to enhance its contribution food security and poverty alleviation.
- Forestry development projects of more than 50 hectares should be subjected to Environmental Impart Assessment (EIA) and should be located on environmental sensitive sites.
- National working group (NWG) on sustainable forest management and forest resources certification should be setup.

Forest and Timber Resources

- Periodic inventories of forest should be conducted to collect data/information useful for the determination of status planning and management of forest resources.
- Management/working plans should be developed and applied to all forest and plantations.
- Adequate measures should be conducted into the processing, utilization and marketing of lesser known/used tree species.

- Rehabilitation of the timber industry should be managed by professionals accompanied by an active tree-planting programme.
- The annual allowable timber should be cultivated and presented.
- An efficient log trading system should be set up to calculated, prescribe and to account for all trees felled as part of annual allowable cut.
- Conduct research into the development of appropriate technology for processing small diameter trees from the natural forest and plantations.
- An aggressive afforestation scheme should be embarked on.
- Lesser known/used species of trees should be aggressively promoted to achieve their acceptability in the global wood market.
- Low impact logging as a means of impacting minimal damage on the forest and its wildlife should be adopted and encouraged.

Forest Industries

- Periodic survey should be conducted to determine the number, type, size, capacity, raw material/labour needs e.t.c of plants, machinery in the forest industries.
- Forest concession contracts should specify the need for reforestation of coupes exploited by the harvesting firm/ individuals.
- The total growing stock harvested per year should respect maximum girth/diameter limits and not exceed the prescribed annual allowable cut.
- The private sector should form partnership and collaborate with stakeholders on issues of supply of forest industrial raw materials.
- Government should provide enabling environment such as to reduce duties on imported inputs in order to increase the profitability of investments in the forest industries.

- As much as possible forest industries should be intergrated so that residues from one process can form input to succeeding down process e.g., particleboard sawdust briquettes etc.
- Timber grading rules should be established and applied.

Non-Wood Forest Resources

These include gums, mushrooms, vegetables, and fruits etc.

- Research should be carried out into development, processing, marketing and utilization of non-wood forest products.
- Farmers and general public should be encouraged to plant and own orchards of indigenous forest fruit trees.
- Homestead and herbal gardens should be established to propagate plants that have medicinal values.
- Management/Working plans should be prepared for sustainable development of non-wood forest products.

Wild Life Resources

These include plants etc.

- Periodic inventory/census of wildlife resources should be taken to determine the status and provide data for sustainable development of wildlife resources in Nigeria.
- Conventions on trade and trafficking on endangered species should be strictly enforced and reviewed from time to time to reflect the status of designated endangered species as revealed by periodic census of wildlife.
- Potentials of wildlife and national parts for enhancing income from tourism should be emphasized.

In addition other long-term prescriptions/polices emphasizing SFM should be put in place:

- Polices requiring corporations/enterprises engaged in large scale farming, mining and oil exploitation and timber and logging should have concession agreements to be an active part in forest include regeneration/reforestation, and rehabilitation in the areas where they carried out their activities before leaving such areas.

- Polices for exploration of cheaper alternative sources of energy to firewood should be pursued.
- Polices geared towards the expansion of high profile but orthodox productive activities centered on forest resources and reforestation.
- Polices of commitment to continuous training and retracing of all cadre of manpower in the act of SFM should be pursued.
- Polices geared towards small scale enterprises, particularly forestry, agriculture and manufacturing to exploit valuable resources related exports opened up by emerging regional trading blocks.
- Policy of unambiguous commitment to all-round capacity building and continuous technological development/innovation in an attempt to keep the products of forestry competitive in global market for sustainable development.
- Policy of utilization of local expertise and development of local governance capacity among host communities in the process of production, transportation and marketing of forest and allied sectors products.
- Introduction of reforestation fund/tax per taxable adult to enable reforestation and check unnecessary deforestation. The fear here is that empowered government agencies might divert the funds to other purpose or even use it as a basis for exploitation of individuals.
- Promoting forestry and economic policies and forestry acts/regulations which ensure most efficient allocations of resources including market based incentives system and appropriate pricing policy.
- Untying of all development assistance, and to recognize the special vulnerabilities of SEM in Nigeria.

- Building consensus with the World Bank, IMF, and other agencies in support of new global framework for enhanced aid and effectiveness of SFM.
- Developing national policies that encourage international institutions to treat forest regeneration in Nigeria as priority sector.
- Developing national policies that support the development of new, creative and economically sound machinery for attracting private investment in the forestry sector.
- Encourage individuals/communities to establish hand own woodlots through subsidized supply for seeding, operation machinery and tax rebates.
- Full development of the idea of social forestry to assist in addressing the nature and level of reforestation required to adequately tackle the problems of deforestation.
- Nigerian governments should treat deforestation as an ecological disaster by setting up a monitoring system for the collection of data on deforestation in order to assist degraded areas within and outside constituted forest areas.
- Purities fines and fail sentences should be prescribed for those felling trees without official permission or starting burn fires.
- Free nurseries should be set up to make trees always available at highly subsidized rates.
- Monitoring tree planting programmes to make them more effective and result oriented.
- State departments/ministry of forestry should be made to throw back a reasonable percentage of forest revenue to assist afforestation/forestation campaign.
- Establish a forum for SFM at which stakeholders discuss, evaluate and co-ordinate efforts to redress forest/environment deforestation. In particular establish opportunities for consultation, partnerships

and listening to the host communities. Such forum should lead to the dialogue with stakeholders implementing of strategies for SFM and poverty reduction.

- Install/programme for genine assistance and global coalition, leading to development of national consensus on the strategies, actions and financing necessary for SFM and poverty reduction programmes in line with the requirements of the united nations forest forum (UNFF) and negotiating framework co-operation agreement with the FAD to support the implementation of national forest programmes in these states.
- Putting in place a perspective plan which emphasizes small scale farming and forestry businesses and policies for improving access to credit and unified extension services that would disseminate transparent information on improved forestry and agriculture with the industrial sector through setting up forestry and agro-processing industries.
- Put in place, national forestry and allied sector perspective plan which should last for 30 years in the first instance. Made up of annual budget within a rolling plan, that should constitute the planning framework.

Finally the management indicators of Nigeria forest seems to be in the direction of mindless pillage of the forest for power and riches by both private businesses, individuals and government at the expense of sustained and environment quality protection. A state of decay and confusion that cannot support economic, agricultural, Industrial, technological and political development or prosperity.

In the 21st century, against this background, the emerging questions are which way forward for Nigeria's forest and wildlife conservation/management and growth potentials realization? Is a re-think necessary, aiming at massive all embracing policy reform? Should it be the British evolutionary reform or the American constitutional reform? Or should Nigeria forest management resign itself to the present pessimistic and non-performing forestry intersectoral linkage architecture? If a combination of global partnership and strategic policy initiatives fail to achieve the derived optional and sustained development what should be the way forward? Should an alternative order be positioned? The debate is open.

REFERENCES

G. E. Ulopong Z, (2003), *An Energy Options in the Industrial and Development Process of the Nation, What Role for Coal,* Bullion 22(4), September/December. Lagos.

Akinsanmi F. A., (1999), *Updating Environmental Information in Nigeria,* A Paper Presented at the 25th Annual Conference of the Forestry Association of Nigeria, Held in Maiduguri, Bornu State, November 7-12, 1999.

Awake, (2003), *Who Will Save the Rainforest?* Benin City: Jehovah Witness Publications, June, 22: 6-8.

'Unianmikogbo Sam O., and Okosoda L. A., (2003), *The Environment and Sustainable Development in Nigeria: Issues for Consideration,* Ekpoma: ..... Publishers.

Egbon Peter C., (2003), *Unpublished Teaching Notes,* Masters in Energy and Petroleum Economics Degree Programme, Delta State University Abraka.

Isimekha 2. A., (2002), *Sustainable Forest Resources Development in Nigeria,* Bullion 26(3): 34-36, September. Abuja: Central Bank of Nigeria.

2

MACROECONOMIC ENVIRONMENT AND AGRICULTURAL SECTOR GROWTH IN NIGERIA

Emmanuel O. Eyo

INTRODUCTION

In Nigeria, the agricultural sector has been invaluable in supporting economic growth and development since 1960. In fact before the discovery and exploration of petroleum, the Nigerian economy depended on funds generated from agricultural export expansion for the development of other sectors of the economy. Due to its important role in nation building, the agricultural sector has continued to be a target of government policies overtime.

Efforts to fortify the Nigerian agricultural sector have a long history. At independence in 1960, the first task of government was the development of high level manpower resources, (Wells 1970). This was in addition to funding agricultural research and extension activities, farm settlement schemes, plantation agriculture, cattle ranches and irrigation project. In 1971, government established the agricultural research council, and introduced the National Accelerated

Food Production Program in 1973. Between 1974 and 1980 government embraced the direct food production policy, introduced the model farms, the Nigerian Grains Production Company, the Nigeria Root Crop Production Company, the Naiwa Mechanized Farm Limited, the Nigerian Dairy Company Limited and the Nigerian Poultry Production Company. Also, government established the Department of Agricultural Cooperatives, provided support to the small farmers and commenced the pilot phases of the Agricultural Development Project, established the Agricultural Project Monitoring and Evaluation Unit to monitor agricultural project of government and established new research institutes. Other programmes introduced to fortify the sector were the Strategic Grains Reserve Program, the Operation Feed the Nation, the River Basin and Rural Development Authorities, the Land Use Act, the Green revolution program, and the Federal Agricultural Coordinating Unit (Eyo 2005). By 1986 government deemphasized the direct food production policy and adopted an integrated approach that combined agricultural with rural development. As a result of this new Agricultural development thrust, government established the Directorate for Food, Road and Rural Infrastructure to construct and maintain feeder road. However, by 1988 government agricultural policy thrust was directed at providing support services while all aspects of agricultural production and marketing including input supply as well as agricultural mechanization were to be handled by the private sector, (Eyo 2005). To enhance the provision of the support services in all parts of the country, government established the Agricultural Development Projects in all state of the federation and established the National Agricultural Land Development Authority. With the growth in the number of government agricultural development programmes, one expected meaningful agricultural output growth, positive change in farm sizes and general development in the sector over time. Unfortunate the index of real agricultural sector GDP shows complete absence of any substantial growth. For example, the growth rate in real GDP of agriculture between

1981 and the year 2005 fluctuated between –13 and 65.13. Table 1 shows the growth rate in real agriculture GDP. According to this table, the growth rate in real GDP agriculture was positive in 1982, between 1987 and 2003 and negative by 1983 to 1986, as well as 2004 and 2005.

Table 1
Growth Rate of Real GDP-Agriculture

Year	*Index of Real GDP Agric*	*Growth Rate*
1981	100.00	
1982	103.85	3.85
1983	96.20	-3.80
1984	90.59	-9.41
1985	100.13	-0.13
1986	98.44	-1.56
1987	136.35	36.45
1988	136.35	36.35
1989	105.30	5.30
1990	118.43	18.43
1991	122.71	22.71
1992	128.07	28.07
1993	132.39	32.39
1994	127.60	27.66
1995	131.19	31.19
1996	137.68	37.68
1997	143.18	43.18
1998	143.76	43.76
1999	143.33	43.76
2000	141.37	41.37
2001	159.08	59.08
2002	163.39	63.39
2003	165.13	65.13
2004	69.52	-30.48
2005	71.79	-28.21

Source: CBN Statistical Bulletin (Various Issues)

On the other hand, the index of real GDP for agriculture fluctuated from 100 in the base year to 98.44 in 1986, from 136.35 in 1987 to 141.37 in the year 2000, increased steadily to 165.13 in the year 2003 and then dropped to 71.79 by the year 2005.

Usually several factors can be said to be responsible for the dismal performance of the Nigerian agricultural sector.

Pre-eminent among these factors is the macroeconomic environment. The macroeconomic environment comprises fiscal, monetary, exchange rate, income and other policies that are used to regulate production activities not only in the agricultural sector but in the other sectors of the economy. Unfortunately, macroeconomic policy outcomes vary greatly depending in part on the policy targets and instruments used, (Agu, 2007). However, several authors (Binswanger, 1989; Kwanashie et al 1998; and Killick 1990) agree that agricultural production marketing and financing decisions are influenced by the macroeconomic environment. More so, Garba, (2000) and Akpokodje (2000) confirm that major macroeconomic policy shifts heighten agricultural policy instability. This paper assesses the macroeconomic environment in Nigeria. and analysis the effect of the macroeconomic policies on agricultural output growth over time.

METHODOLOGY

The Study Area

The study area is Nigeria. It has an area of 923,769 square kilometers and a population of about 88 million people. It is bounded on the West by the republic of Benin and the republic of Niger; on the East by the republic of Cameroon; on the north by Niger and Chad republics and on the South by the Gulf of Guinea. The climate is equatorial and semi-equatorial. There are two seasons; the wet and the dry season and agriculture is a major employer of labour.

Source of Data and Data Analysis

This study uses principally secondary data obtained from the Central Bank of Nigeria the Nigerian National Planning Commission. Existing literature indicate that prices, government expenditure in agriculture, volume of credit to the agricultural sector, nominal interest rate and exchange rate which are indicators of monetary, exchange rate and price policies determine activities in the agricultural sector, (Fosu, 1992; and Garba, 2000). Consequently, in this study, data on exchange rate, nominal interest rate, world prices, credit to the agricultural sector, and government expenditure on the agricultural sector, inflation rate, and foreign private investment in agriculture were obtained between 1970 and 2005 and used as indicators of the macroeconomic environment. The method of data analysis is the multiple regression analytical technique (Ordinary Least Square Procedure). The problem of autocorrelation and multicollinearity were aborted by the use of relevant econometric tests, namely; the Durbin Watson test, the *F*-test, the *t*-test and the coefficient of multiple determination.

The structural response function used by Fosu (1992), Amin (1995), Kwanashie et.al (1998), and Umoh (2003) were adapted and used in this study.

Consequently, the model postulates that the index of agricultural production is a function of the indicators of the macro economic environment, namely; foreign exchange rate (*F*), nominal interest rate (*N*), credit to the agricultural sector (*A*), world prices (*W*), government expenditure on agriculture (*G*), rate of inflation (*L*), foreign private investment in agriculture (*P*) and time (*T*).

Consequently, $Y = f(F, N, A, W, G, L, P, T)$

The implicit form of the model is given as:

$$Y = a + b_1F + b_2N + b_3A + b_4W + b_5G + b_6L + b_7P + b_8T + u$$

Where

Y = Index of agricultural production

F = Foreign exchange Rate

N = Nominal Interest rate on loans

A = Credit to Agricultural sector *n* Naira.

W = World Prices of agricultural produce in Naira per tone

G = Government Expenditure in Naira

L = Inflation Rate in Percentage

P = Foreign private Investment in Naira

T = Time trend representing technological change overtime.

a = intercept; $b_1 - b_8$ = coefficients, u = stochastic disturbance term.

RESULTS AND DISCUSSION

Many macroeconomic policy strategies have been tried in Nigeria in an attempt to meet the objectives of diversifying the production and support base away from dependence on petroleum, among other things. In particular several efforts are identifiable in the area of exchange rate, fiscal, income and monetary policies. For instance, the exchange rate policy adopted as many as three strategies, namely; pegging of the Naira exchange rate to the Dollar; the dual exchange rate regime of 1995 to 1999 and allowing the exchange rate to be determined by forces of demand and supply. The fiscal policy strategies adopted in Nigeria so far included the introduction of value added tax to replace sales tax, the abolition of excise duties, regulation of custom and excise tariff from time to time, and the retention of tariff concession granted to the transport sector, among others. Similarly, the strategies of the income policy have included wage freeze of 1976/77, 7% wage increase of 1977/78, increase in minimum wage and allowances of 1980, among others. Apart from these, the Structural Adjustment Policy (SAP) introduced in 1986 set in motion a stream of reforms that completely reversed the effect of the macroeconomic environment on many sectors of the economy. The reforms of the SAP used several policy instruments. First, there was a recall of Naira counterpart of outstanding external payment arrears to the Central Bank of Nigeria, second, there was increase in permissible credit expansion in 1988, third, there was increase in commercial bank cash ratio requirement which started in 1989, fourth, there was also the introduction of cash ratio requirements for merchant banks and raising of the liquidity ratio of commercial banks from 25 to 30%, fifth, there was the lifting of credit ceiling for healthy banks and increase of minimum cash reserve requirements of banks from 3 to 6% in September 1992. On the whole, during SAP, efforts were made to strengthen bank supervision, and new banking standards were introduced minimum capital was increased, special interest on agricultural loans was abolished and subsidy on agricultural inputs were removed. In the post SAP period, more macroeconomic policies were introduced which brought about more financial reforms. As a consequence of the new policies a number of micro credit programmes have been introduced, government merged the Nigeria Agricultural Cooperative bank, the Peoples bank and the Family Economic Advancement Program to form the Nigerian Agricultural Cooperative and Rural Development Bank in the year 2000. By the year 2004 the minimum capitalization of banks was raised to twenty five billion Naira, all public sector funds were withdrawn from commercial banks, and mergers as well as acquisitions were allowed as a means of consolidation of banking institutions. Also, community banks were converted to micro finance banks, and their minimum capitalization was raised to a minimum of twenty million Naira of shareholders fund unimpaired by losses (Imala, 2005).

Table 2
Percentage of Agriculture in Total Export

Year	*% Agric export*	*Other export*
1982	2,03	97.97
1983	3.58	96.42
1984	2.62	97.38
1985	2.08	97.92
1986	4.90	95.10
1987	2.95	97.05
1988	6.35	93.65
1989	2.11	97.89
1990	1.71	98.29
1991	1.31	98.69
1992	1.20	98.80
1993	1.7	98.30
Mean Exchange	2.71	97.29

Source: CBN Statistical Bulletin (Various Issues)

Unfortunately, existing data suggest that the macroeconomic environment created by the macroeconomic policy strategies were not agricultural sector friendly. For example, the exchange rate regime has not resulted in any meaningful export of agricultural produce over time. For example, Table 2 shows that between 1982 and 1993 agricultural export was only 2.71 per cent of the total export, on the average; the yearly percentage of agricultural output in the entire export was highest in 1988 (6.3%) and lowest in 1992 (1.2%).

The proportion of agricultural export in the entire export increased from 2.02% in 1982 to 3.58% in 1983, and then declined to 2.08% by 1985. In 1986 when SAP was introduced the percentage of agriculture in the entire Nigerian export increased to 4.90% and then gyrated to 6.35% by 1988 and then declined steadily to 1.20% and 1.70% in 1992 and 1993, respectively.

Table 3
Maturity Structure of Commercial Bank Loans (1980-1996)

Year	*Percentage of loans that mature in one year*	*Percentage of loan that mature after five years*
1980	78.38	4.86
1981	68.59	6.01
1982	78.70	5.70
1983	81.40	2.00
1984	79.36	5.51
1985	82.08	5.00
1986	81.96	4.70
1987	80.82	4.50
1988	83.31	3.90
1989	46.92	2.33
1990	82.01	5.30
1991	82.31	4.87
1992	82.66	5.84
1993	83.11	5.33
1994	84.40	4.59
1995	83.96	4.72
1996	83.00	5.00

Source: CBN Stat. Bulletin 2005

More so, the macroeconomic environment culminates in the financial sector giving mostly short term loans. This is evident from the distribution of loan maturities in the economy. According, to Table 3, over 68% of the commercial bank loans had loan maturities of one year or less between

1980 and 1988. By 1989, only 46.92% of the loans had maturities of one year and less, but beginning 1990, over 80 per cent of commercial bank loan to the economy had maturity of within one year. Loans with short maturities are mostly operating loans, which are inadequate for the acquisition of intermediate and fixed assets that are invaluable in improving the capital base of the sector.

Besides, an assessment of credit to the agricultural sector in Table 4 shows that between 1970 and year 2005 the real credit to agriculture was consistently low.

Table 4
Mean Volume of Credit to Agricultural Sector in Real Terms

Year	Mean volume of credit to *agricultural (Millions naira)*
1970-1974	1.25
1975-1999	4.91
1980-1984	12.34
1985-1989	16.21
1990-1994	14.72
1995-1999	15.45
2000-2005	42.45

Source: CBN Statistical Bulletin (Various Issues)

It was 1.25 million on the average between 1970 and 1974, 4.19 million Naira on the average in 1975-1979; 12.34 million on the average from 1980 to 1984; 16.21 million Naira in 1985 to 1980; it was 14.72 million on the average in 1990-1994; 15.45 on the average in 1995-1999 and moved to 42.45 million in the year 2000-2005.

Effect of the Macro Economic Policies on Agricultural Output Growth

The effect of these macroeconomic policy strategies on the agricultural sector is captured in the following regression equation.

$$Y = 70.06 - 0.14\ ER + 1.11\ NR + 9.89\ AC + 4.23E - 03WP + 9.02\ E - 05GE$$

$$t = 6.38^{*} \quad t = -1.86^{***} \quad t = 2.52^{**} \quad t = 0.19 \quad t = 2.841^{*}$$

$$t = 2.241^{**}$$

$$- 0.155 \text{ INFR} + 3.38 \text{ E} - 02 \text{ FPN} - 0.45 \text{ T}$$

$$t = -1.26 \quad t = 3.672^{*} \quad t = 0.90$$

$$R^2 = 96\%; \text{ Adjusted } R^2 = 94\%$$

ST. Error 9.68; F. Value 72.043* Durbin Watson = 1.741.

This result shows that the explanatory variables explained 94 per cent of the total variables in the index of agricultural production, all the explanatory variables were collectively relevant as evident in the *F*-statistic being significant 1% level and the Durbin Watson statistic is plausible. However, credit to the agricultural sector, inflation rate and technology changes were not significant in their effect on the index of agricultural output whereas the exchange rate regime, nominal interest rate, world prices, government expenditure and foreign private investment in agricultural were significant in their effect on the index of agricultural output. It is also clear from the regression analysis that exchange rate is negatively related to the index of agricultural production. Consequently, at favorable exchange rates, more agricultural outputs would be produced.

Against the apriori expectation, the nominal interest rate is positively related to the index of agricultural production. This implies that at higher nominal interest rate, more credit facilities are made available to the operators of the Nigerian agricultural sector; but at lower nominal interest rate, credit facilities are no more widely available. The index of agricultural production is also positively related to world prices of Nigerian major agricultural commodities. This implies that better world prices would enhance agricultural output growth. Similarly, the index of agricultural production was positively related to government expenditure on agriculture and had no significant effect on agricultural output growth. Consequently, more government investment would greatly enhance agricultural output growth. The index of agricultural production is negatively related to the level of inflation, implying that the higher the rate of inflation the lower the index of agricultural production. Foreign private investment in agriculture is positively related to the index of agricultural production whereas technology is inversely related. This confirms that the level of technology used in the Nigerian agricultural sector has been grossly inadequate.

CONCLUSION

An intuitive appeal of the salient points of the theories of economic growth clearly indicates that sustained growth in per capita agricultural output is possible through capital accumulation and technical progress, among other things. More so, the theories of agricultural development, particularly, the high pay-off input theory emphasizes that sustained agricultural output growth can be achieved by applying high pay-off inputs as well as new technologies like improved planting materials, fertilizer, irrigation facilities, improved livestock species and motorized equipment. This research confirms that the technological progress is limiting, as well as credit to the agricultural sector. Invariably the macroeconomic environment has not been able to support the operators of the agricultural sector to acquire high pay-off inputs that are invaluable in improving the capital base of the agricultural sector. Intermediate and fixed assets make up the capital base of the farm business and contribute significantly to the size of the farming operation. Therefore their acquisition is important to the growth and development of the agricultural sector and access to external financing would be invaluable in accelerating the growth process. In view of these, macroeconomic policies that enhance favorable exchange rates, make agricultural credit widely available at low interest rate, reduce the rate of inflation, increase foreign private investment in agriculture, would not only fortify government investment in the sector but would be invaluable in supporting agricultural output growth in Nigeria.

REFERENCES

Akpokodje G. (2000), The Effect of Export Earnings Fluctuations on Capital Formation in Nigeria, *African Economic Research Consortium*, Research Paper RP103 October 2000, Nairobi, Kenya.

Agu C. (2007), What Does the Central Bank of Nigeria Target? An Analysis of Monetary Policy Reaction Function in Nigeria, *Final Report Submitted to the African Economic Research Consortium*, Nairobi, Kenya, December 2007.

Amin A. A. (1996), The Effect of Exchange Rate Policy on Cameroon's Agricultural Competitiveness, AERC Research Paper 42. Initiative Publishers. Nairobi,

Binswanger H. (1989), The Policy Response of Agriculture: In: Stanley F. and D. de Tray (Eds), *Proceedings of the World Bank Conference on Development Economics*, Washington DC, 121-53.

CBN (2005), Statistical Bulletin, 16, December.

CBN (1994), Statistical Bulletin, 5(2), December.

Eyo E. O. (2005), Agricultural Development in Nigeria: Plan, Policies and Programmes. Best Print Business Press, Uyo-Nigeria.

Fosu K. F. (1992), The Real Exchange Rate and Ghanas Agricultural Export AERC Research Paper 9. Kenya, Nairobi.

Garba P. K. (2000), An Analysis of the Implementation and Stability of Nigerian Agricultural Policies, 1970-1993, *African Economic Research Consortium*, Research Paper 101, Nairobi, Kenya.

Killick T. (1990), Markets and Government in Agricultural and Industrial Adjustment, ODI Working Paper 3, London.

Kwanashie M. Garba A., and I. Ajilima, (1997), Policy Modelling in Agriculture: Testing the Response of Agriculture to Adjustment Policies in Nigeria, AERC Research Paper RP 57.

Umoh G. S. (2003), Nigeria: Macroeconomic Environment and the Perennial Crops, *Global Journal of Agricultural Science*, 2(1): 53-61.

Wells J. C. (1970), Issues in Agricultural Development Policy During 1962-68, Nigerian Development Plan, In *Growth and Development of the Nigerian Economy*, Edited by Carl K. Eicher and Liedhorim, Michigan State University Press.

3

IMPACT OF WTO ON AGRICULTURAL MARKETING

R. B. Dhande & G. D. Kharat

INTRODUCTION

The world agricultural trading system has been experienced a tremendous change during the second half of the 20th century. The tread on agriculture was signed at the ministerial meet on 15th April 1994 by 125 nations and the agreements of agriculture (AOA) become effective from 1st Jan 1995 with the establishment of WTO.

The WTO agreements are a milestone in the development of International trade in agricultural commodities; agriculture has been brought under international discipline. It is very difficult to foresee that exact outcome of WTO.

Agriculture marketing was tilled recently, not fully accepted as an essential element in agricultural development in the countries of Asia Agricultural marketing occupies as fairly low place in agricultural development policies of developing countries. Agricultural development means not only enough to produce a crop but also we must see it to be satisfactorily marketed. The present paper has been described the need for better agricultural marketing. Foreign customers demand quality products. The challenges left before the agriculturists providing them an opportunity for making some heroic efforts to capture the world market. Traditionally, plantation & fiber crops occupied primary position in agricultural exports of our nation.

OBJECTIVES OF THE STUDY

The present study is an attempt to:

1. Examining an impact of AOA on Indian Economy.
2. Identifying the need for better agricultural marketing.

HYPOTHESIS

The marketing (both domestic and export) of agriculture is a decisive issue & hence requires an intensive strategy with structural & infrastructural facilities for the development of well built marketing system.

METHODOLOGY

The methodology is designed with an intention to make a critical appraisal of the present day agriculture & identify the challenges. The study is based on secondary source of data.

AGREEMENT OF AGRICULTURE (AOA)

Agreement of agriculture (AOA) is related to reforms in agricultural trade over time. Accordingly member countries have to formulate market oriented policies. The commodities under AOA are to be fulfilled through:

(a) Providing market access

(b) Reducing domestic support

(c) Containing export subsidies

The member countries have to provide market for agricultural products to the extent of 3% of domestic consumption by end of implementation period i.e. 2000 & 5% latter on. In order to move towards free trade, member countries have to reduce tariffs. Developed countries have to reduce import tariffs by 36% for all products (at least 15% per product) whereas developing countries 24% of all products (at least 10% per product) from present level. Under WTO these have been replaced by tariffs that provide equivalent levels of protection. If the domestic prices were 25% higher than the prices prevailing in International Market, now import tariff called tariffication could be 25%.

As a part of AOA, subsidies & other form of support to agriculture including minimum support prices & procurement prices have to be used less intensively by the member countries, to promote agriculture sector in their respective economies.

WTO categorized the domestic support in terms of trade distorting called as 'amber box', 'green box' & 'blue box'.

'Amber box' includes the domestic support policy induce farmers to produce more than the domestic requirement. The presence of over production forces the Govt. to provide subsidies to clear it out. The trade distorting domestic support is measured in terms of Aggregate Measure of Support (AMS). It means the difference between minimum support price & external price quoted in the International Market.

'Green box' includes minimal trade distorting domestic support toward scientific research, disease control, development of infrastructure & to ensure food security (to maintain PDS). It also includes the direct payment made by Govt. to sustain incomes of farmers on the one hand & other to discourage farmers to cultivate certain crops & to limit the production through declaring crop holiday.

The domestic support in the form of direct income payments to farmers is called as 'Blue box'. The members can continue other domestic support on small scale, called as 'Special & differential box'.

The other major issue related to agriculture is containing export subsidies. The member countries have to contain the quantity of subsidized export as well as the amount earmarked towards exported subsidy. The developed and developing countries agreed to cut the subsidized quantity & amount of money they spend on export subsidies.

IMPACT OF AOA ON INDIAN ECONOMY

Agreement of Agriculture is biased in-favor of developed nations. Provision of marked access clause adversely affects the developing nations particularly India. It is fact that, there is huge market for agricultural products in developing nations and at the same time, there is surplus food production in developed nations. In the guise of AOA & its market access clause, the developed nations will dump their surplus food grains in developing countries.

MARKET ACCESS CLAUSE

India has reached to stage of self sufficiency in food grains. We do not need impact of food grains. But under this clause, we have to provide market access for agricultural products to the extent of 3% by spending precious foreign exchange. AOA has been structured in such away to deny increased market access to developing nations. India could not derive any advantage from market access clause. Being a member of WTO & a party of AOA, India has to fulfill its commitments by allowing impacts of agricultural products in the presence of adequate domestic supply of agricultural products. The import of agricultural products even in the presence of glut of output has adverse impact on Indian economy in terms of fall in prices of agricultural products & income levels of farm community.

DOMESTIC SUPPORT

The second major issue under AOA is related to containing domestic support especially subsidies to agriculture sector. AOA stipulated that total subsidies to agriculture should not exceed 10% of the value of GDP of agriculture sector.

The developed nations providing subsidies to agriculture well beyond stipulated limits. USA spends more than 1/3rd of GDP from agriculture, Japan 1/4th of the GDP from agriculture, Canada & European Union around 10% of GDP from agriculture towards green box subsidies.

In India, the amount of subsidies extended to agriculture sector constitutes less than 10% (near about 6%) of GDP from agriculture. In our economy the adverse impact of weather, defective seeds & rise in fertilizers prices on overall crop yields. Under these conditions, the government has to provide subsidies to protect farm sector & to ensure food security. If the Indian Government under AOA in near future, agricultural output may fall due to lack of incentives to farmers & focus our millions of people to look at developed countries to satisfy their hunger. This is the final outcome that the developed nations aim to achieve under AOA. Though we are a party of AOA, we should not fall in the trap of developed nations.

EXPORT SUBSIDIES

Regulation of export subsidy is a major factor that adversely affects the performance of agriculture sector. Under AOA, the government has to regulate export subsidy in terms of duty drawbacks, cash compensatory support, etc. The aim of developed nations, under the guise of AOA, is to make the exports of agricultural products from developing nations less competitive in the international market through regulation of export support. The provision of market access, containing domestic support & regulation of export subsidies clearly indicates that, developed nations have to clever vision in terms of expecting a huge market for their agricultural products in developing economies. According to the clause of "technical barriers to trade "also prepared in favor of developed nations. Under this clause, developed nations can effectively prevent imports of agriculture & allied product from India, on health grounds, environmental issues & citing violation of human rights in the form of employing child labor.

NEED FOR BETTER AGRICULTURAL MARKETING

Agricultural marketing plays an important role not only in stimulating production & consumption but in accelerating pace of economic development. Marketing of agricultural produce is complex phenomenon. It requires special treatment at all levels with government & institutional support.

Indian farmers are poor and illiterate; consequently do not acknowledge the new technology in pre harvest operations. All sort of pre- harvest operations are not market oriented & sustainable. Quality of production also annoyed due to natural calamities & climatic conditions. Crop protections measures are becoming costly. Cost of input also remains high. There is no guarantee of prices to be paid for farmers produce. The producer's share in consumer rupee is tremendously less in India. All the lacunars deserve to have Institutional and infrastructural facilities to be made available to the farmers by government authorities. The government has taken various steps to improve the conditions of the agricultural marketing, but these steps are not meeting the required demands. In the wake of globalization, the farmers have to face lot of challenges ahead to market his product. Some of the initiatives needed for tomorrow's better marketing are as follows:

1. Expansion of Co-operative marketing & processing:

   Expansion of co-operative marketing facilities for all commodities assures reasonable prices so as to remove all exploitation of the farmers by the middlemen. A network co-operative at the national level, state level & primary level needed to help farm producers to get best markets & best prices for their product. These markets have their own storage & warehousing facilities which results in better control over the flow of supplies & influence the prices.

2. Electronic Trading:

   It may facilitate the direct interaction between buyers & sellers. Agricultural market committee can also set up the special information kiosks for this purpose. The issue in the concept calls for the dual role of farmers as producers & seller. With the present infrastructural facilities can we justify it?

3. Information Network System:

   The ministry of agriculture has recently implemented an I T based central sector scheme to link 710 important regulated markets spread all over the country including the state Agricultural Marketing

Board & Directories. Under this scheme, an agricultural information system network is being set up to computerize 27 major commodities with online data of daily market arrivals, prices, storages, dispatches with destination cost, sold & unsold stocks, infrastructural facilities, etc with this advances the world has become a small place and a larger market at one can go.

4. Forward & Future Markets:

   In the wake of WTO, the government has withdrawn its intervention from commodity sector to ensure that the demand & supply forces will determine the market prices in the new economic environment. The Forward Markets Commission (FMC) performs the functions of advisory, monitoring, supervise & regulating in future and forward trading. In the century, currently future controls are traded for nine commodities in twenty commodity exchange. The linkage between spot & future markets seem to be poor due to domination of speculators. The government therefore has to continue its efforts to strengthen traders. Efforts have also to be made to extend the role of exchange in other traders and other services like warehousing, financing, investments and market intelligence which could improve the existence of the future markets.

5. Creation of Cold Storage Facility:

   In India, presently, the total warehousing in both urban & rural areas which accounts to 35 million tones (2000-2001) capacity. Now, the need of the hour ahead is creation of cold storage facilities to growing demands of fruits & vegetables, where we have to convert our comparative advantage in the production & process of fruits & vegetables, to competitive advantage which is still untapped potential. Very merger efforts have been made in a form location specific Agri-Export Zone [AEZ] in this area which could not sufficient. It may be noted that plantation crops like tea, coffee, spices, coconut, cashew, dairy products require storage facilities. Efforts will have to create crop-specific storage capacity during the 11th five year plan.

6. Linkage Of roads:
   As compared to railway, the road transport system has definite advantages. Road development in India was neglected in past years for various reasons. In the first place, central & state governments did not appreciate the importance of developing the road system. During the next three plans, the planning commission provided for a total outlay of Rs.8600 corers on road programs & for road transport. One of the most ambitious projects launched in independent India is the National Highway Development Project (NHDP), which is expected to go a long way in promoting economic development of the country & integrating remote regions with the main stream economic activity and ensuring connectively of the rural roads.

7. Private Investment:
   The decline in the share of Investment in Agriculture as percentage of GDP has been an area of continued concern. This decline was attributed mainly to near stagnation or fail of public investment in agriculture. For the development of agricultural marketing, private sector must invest their share in agriculture in the coming years.

8. Price Support Policy:
   Food grain marketing has been a subject of considerable debate in the recent past. The government intervention in food grain market in case of rice & wheat in a farm price support, buffer stocking & PDS helped a lot in reaching the rural poor. Today food security is one of the prime importances of the nation. There are only 24 agricultural

commodities which are covered under minimum price support policy & have regulated markets throughout the country. There are several other commodities which of high economic significance that need urgent coverage through comprehensive price stabilization as well as establishment of regulated markets in the respective agro climatic regions. There is a need for policy which should have high market intervention and it should increase the operational efficiency of purchasing agencies in the context of cost efficient purchases vis-à-vis competitive sales as to avoid or reduce loss. There is a need for bringing uniformity in the state level tax structured for better efficiency.

The infrastructure under government ownership & management has, however, proved thoroughly inefficient and corrupt. The demand for infrastructure facilities & services has always out paced supply, besides the quality of the existing supply is extremely poor. The widening gap between demand and supply of infrastructure & the extremely poor quality of the existing supply raises important questions concerning the sustainability of economic growth of the country in the coming years.

After considering the above issues, it is sure that we are able to develop well built marketing system.

REFERENCES

Acharya S. S., and Agrawal N. L. (2004), "Agricultural Marketing in India," Oxford and IBH Publishing Co. Pvt. Ltd.

Arunachalam R. (2002), "An Analysis of India's Steucture and Direction of Foreign Trade," *Agricultural Growth & Economic Reforms*, 215-225.

Mercia Selva Malar S. (2005), "WTO in Support of Agriculture Trade," *Kisan World*, 63.

Prasad (2005), "Encyclopedia of Agricultural Marketing", Vol XI.

FAO. (2005), The State of Food and Agriculture, Rome, Italy.

Josling T., and S. Tangermann (1999), Implementation of the WTO Agreement on Agriculture and Developments for Next Round of Negotiations, *European Review of Agricultural Economics*, 26(3): 371-388.

Dr. A. Nakula Reddy (2006), Multilateral Trade Agreement on Agriculture - Impact on Indian Economy, Osmania, *Journal of International Business Studies*, July-December 2006: 51-56.

4

ENVIRONMENT MANAGEMENT ACCOUNTING [EMA]: AN OVERVIEW

M. Sumathy

INTRODUCTION

Environmental costs and obligations are significantly growing and continue to grow as the world becomes more environmentally conscious. Public corporations are being held more responsible and accountable to be good environmental citizens. A number of environmental legislations have been enacted to address the emergence of a worldwide "green movement."

In some cases in years past, environmental issues were virtually ignored by both corporations and individuals. Hazardous waste and other such items were considered a cost of a growing economy. Times have changed as people now realize the effects of waste products that potentially could damage parts of the environment. Many people are willing to pay more for a product that is environmentally friendly. Many companies are now interested in being "green," as many investors place a high value on environmental responsibility. Some corporations have had to pay to clean up their past environmentally "unfriendly" behavior. However, most firms have established good reputations as environmentally "friendly."

Conceptual frame work of environment management accounting has been describe in "Rig-Veda", the pioneer religion theorem of Hindu Philosophy as according "sky is like a father, earth is like a mother, a space as their children." Thus Rig-Veda reflects that environment is to be valued like parents and love like children. The concept of environment accounting and auditing may be new for western accounting thinkers. There may be conflicts between understanding and implementation. Few notes started its relevance for country. Such as establishment of new accounting standards concerning to environment management accounting in India. Secondly it should require to include into specialized accounting education syllabus and to create awareness of importance of environmental accounting to the accounting practitioners for serving its exact purpose.

EMA: Background and Overview

Environment Management Accounting (EMA) is the identification, collection, estimation, analysis, internal reporting, and use of materials and energy flow information, environmental cost information, and other cost information for both conventional and environmental decision-making within an organization. For companies that have the goals of saving money, especially environmental costs, and reducing environmental impacts, EMA provides essential information for meeting those goals. Key points to note are:

EMA focuses on costs internal to the company; EMA does not include external costs to individuals, society, or the environment for which a company is not legally held responsible.

EMA places particular emphasis on accounting for environmental costs.

EMA also encompasses explicit information on physical flows and fates of materials and energy.

EMA information can be used for most types of management activity or decision-making within an organization, but is particularly useful for proactive environmental management activities.

EMA is not merely one environmental management tool among many - rather, EMA is a broad set of principles and approaches that provides the materials/energy flow and cost data critical to the success of many environmental management activities. Terms or tools such as full cost accounting, total cost assessment, cost accounting, materials accounting, life cycle assessment, life cycle costing, and activity based costing are associated with EMA.

EMA: Reasons for Change

EMA researchers and proponents recognize the limitations of conventional management accounting approaches for management activities and decisions involving significant environmental costs and/or significant environmental consequences/impacts. The unintentional "hiding" of many environmental costs in overhead accounts Inaccurate allocation of environmental costs from overhead accounts back to processes, products, and process lines Inaccurate characterization of environmental costs as "fixed" when they may actually be variable (or vice-versa) and The actual lack of inclusion of relevant and significant environmental costs in the accounting records at all.

EMA BRINGS MANY POTENTIAL BENEFITS TO INDUSTRY INCLUDING

The ability to more accurately track and manage the use and flows of energy and materials, including pollution/waste volumes, types, and fate.

The ability to more accurately identify, estimates, allocates, and manage/reduce costs, particularly environmental types of costs More accurate and comprehensive information to support the establishment of and participation in voluntary, cost-effective programs to improve environmental performance More accurate and comprehensive information for the measurement and reporting of environmental performance, thus improving company image with stakeholders, such as customers, local communities, employees, government, and finance providers.

Environment Management Accounting: Implementing EMA

EMA provides two types of data to assist companies in making decisions about capital investment, product/process costing, strategic planning, or other business opportunities. One type is cost information; the other type is physical flow information (e.g., raw materials use and waste generation rates). Companies may approach the data collection in different ways. For example, some companies may simply conduct a more careful examination of existing cost data and combine it with existing materials flow knowledge to make more informed decisions.

Other companies may do actual mass balances on their processes to better understand previously hidden losses. Some examples of physical flow information include: quantity of chemical brought on-site, quantity of chemical produced on-site, quantity of chemical consumed in the manufacturing process, q quantity of chemical in the product, quantity of chemical in the waste, water use, wastewater generation, and energy use.

Surveys of Industry Practices

The surveys of current industry practices in environment management accountings of major corporations within a variety of industries and provides a perspective of the current state of the art of environmental management accounting, as well as some important differences in organizational approaches. Here the reporting, data handling, and follow-up processes of selected leading environmental management accounting programs were analyzed in-depth to develop perspectives on the diversity in audit reporting techniques, formats, and approaches to provide clear and appropriate disclosure of audit findings. In addition, we have informally surveyed hundreds of companies at our various audit seminars and workshops.

External Environmental Accounting for the Board of Directors

Working for a committee of the boards of directors of several major U.S. companies, serve as external auditors and conduct an independent environmental verification program. As part of these client assignments, the corporation to further develops and strengthens its internal EHS, and product safety audit program. The boards of directors find our assistance to be very valuable to them in assuring compliance.

Audits in Highly Sensitive Situations

Working for outside counsel in dozens of highly sensitive situations, we conduct audits of numerous facilities that pose significant environmental problems. In one such assignment, a team of audit experts conducted several field visits to independently determine compliance with national, state, and local environmental regulations; assess whether facility operations and practices pose potentially serious health or safety risks; and determine conformance with corporate policies and procedures.

Evaluation of Facility-Level Environmental Compliance

For a large international manufacturer of industrial/ agricultural equipment, we developed a program to evaluate the environmental compliance and management systems in place at the facility level. The purpose of the assessments was to identify those areas where environmental risks had not been identified or had not been adequately addressed and to evaluate the management systems in place to direct environmental issues. We provided two to six team members with project-and industry-specific expertise for each assessment. The assessments involved interviews with management, engineering, maintenance, and operational personnel; inspections of environmental activities and review of environmental documentation and programs

REFERENCES

Abbott W. F., and Monsen R. J., (1979), On the Measurement of Corporate Social Responsibility: Self Reported Disclosure as a

Method of Measuring Corporate Social Involvement Academy of Management Journal, September.

Alhashim D., (1977), Social Accounting in Egypt, *International Journal of Accounting*, Spring.

American Accounting Association, (1971), Report of the Committee on Accounting for Social Performance Accounting Review.

Freedman M., and Stagliano A., (1992), European Unification, Accounting Harmonization and Social Disclosures, *International Journal of Accounting*.

Gray R., and others, (1997), Struggling with the Praxis of Social Accounting: Stakeholders, Accountability, Audits and Procedures Accounting, *Auditing and Accountability Journal*.

Guthrie J., and Parker L., (1989), Corporate Social Reporting: a Rebuttal of Legitimacy Theory Accounting and Business Research Autumn

P. C. Mishra, N. Behera, B. K. Senapati, and B. C., (1995), Guru Advances in Ecology and Environmental Sciences, Vedams e-Books (P) Ltd., New Delhi.

APPENDIX 1

ENVIRONMENTAL ACCOUNTING PRACTICES OF LISTED COMPANIES IN JAPAN

The number of Japanese corporations, which publish environmental reports, has been increasing very rapidly. According to the "A Survey of Environmentally Corporate Behavior" [Ministry of the Environment (2001a)], the proportion of listed corporations surveyed which disclosed environmental information showed a rising trend from 35.7 per cent (1998) to 40.9 per cent (1999) to 51.0 per cent (2000). Out of these companies the proportion of those which published environmental reports also increased from 30.9 per cent (1998) to 37.3 per cent (1999) to 45.9 per cent (2000). This sort of trend is likely to increase further, judging from the publication of "Environmental Reports Guidelines (Fiscal 2000)" by the Ministry of the Environment (MOE) in February 2001 and the "Environmental Reporting Guideline for Stakeholders" by the Ministry of Economy, Trade and Industry (METI) in June 2001.

The number of companies, which disclose environmental accounting information in their environmental reports, is also on the increase. During the first half of the 1990s when the word "environmental accounting" was not in general use, only a handful of corporations measured environmental costs. However, according to the Moe's survey (2001a), out of the above-mentioned listed corporations which replied that they disclosed environmental information, the proportion which disclosed environmental accounting information showed a steeply-rising trend from 10.4 per cent (1998) to 20.9 per cent (1999) to 27.0 per cent (2000). Concerning the question on the introduction of environmental accounting, 17.3 per cent replied that they had already introduced it, while 34.2 per cent replied that they were considering its introduction. These trends were obviously influenced by the environmental accounting guideline published by the Environmental Agency (now the Ministry of Environment: MOE) in May 2000. The draft guideline was published in 1999. Furthermore, both of the Moe's and the METI's environmental reporting guidelines recommended environmental accounting information disclosures in the environmental reports Therefore, more and more companies are expected to introduce and publish environmental accounting.

Although such guidelines are likely to have a considerable influence on environmental accounting and reporting practice, they are not mandatory rules, but voluntary. The methods and procedures for environmental accounting in the Moe's guideline are quite flexible and even ambiguous. The guideline leaves much discretion to companies. This means that how and to what extent the guideline influence environmental accounting practice becomes an important research issue. The object of this study is twofold: to clarify the special characteristics of Japanese environmental accounting practice by examining the environmental accounting information disclosure by Japanese corporations; and to analyze the influence on Japanese corporations by the MOE environmental accounting guideline. Before examining these issues, some main governmental initiatives on environmental accounting and previous studies on Japanese environmental accounting practices are briefly studied.

APPENDIX 2

GOVERNANCE OF ENVIRONMENTAL ACCOUNTING PRACTICES IN USA

Industry accountants must ensure that their firms are in compliance with accounting standards, including those pertaining to environmental issues. Several rules governing environmental disclosures have been developed in USA. At the highest level of generally accepted accounting principles (GAAP) are the Statements of Financial Accounting Standards (SFAS), which are issued by the Financial Accounting Standards Board (FASB). Some of the key accounting pronouncements that regard environmental reporting is the following:

1. SFAS No. 5 states under "Accounting for Contingencies" that a liability should be recognized in the financial statements if a loss is probable and the amount is estimable. If the loss amount is not estimable which is often the case, the contingency must be described in the footnotes to the financial statements.
2. ETIF Issue No. 90-8 required that all environmental contamination costs be expensed as incurred unless costs extend the life or increase capacity of the property, costs mitigate or prevent future environmental contamination (that would otherwise occur), or costs are incurred to prepare a property for sale.
3. ETIF Issue No. 93-5 concluded that an environmental liability must be evaluated independently from any potential claim for recovery. This recovery claim can reduce the liability only if it is probable. Securities Exchange Commission (SEC) standards state that it is appropriate to net the asset and liability if the asset's recovery is recognized as probable. The asset and liability should be disclosed in the notes to the financial statements. The SEC approves discounting of liabilities to their present value.

4. AICPA Statement of Position (SOP) 96-1, "Environmental Remediation Liabilities," covers auditing and accounting topics dealing with environmental issues. It details the responsibilities of corporations involved in environmental cleanup, and responsibilities of corporations to avoid environmental destruction.

5

ENVIRONMENT VS DEVELOPMENT: CAN WE HAVE EQUILIBRIUM?

Rameshwari Pandya

INTRODUCTION

The man on the street does not feel "bothered, directly and personally, by farmer suicides and melting Himalaya glaciers", then let us understand well that no amount of structural or policy changes can truly help us face these trials. It will take large number of courageous, creative and caring citizens who feel personally responsible for the state of society and ecology, and are charged with the agenda for transformation.

There are two over arching narratives. The first narrative is of "Emergent India"- a nation growing in confidence and willing itself to become an economic powerhouse. The second theme is that of the "other India" where socio-economic development still remains a distant reality. Paraphrasing a renowned journalist, it is the story of booming sensex and of a hundred thousand farmer suicides. There are several challenges which we are facing today. The most important challenge is environmental. It is estimated that, given the speed of glacier melt, the major rivers will become seasonal,

Rising sea levels will threaten our coastal communities and monsoon will become more erratic. Global warming is reality and will begin to impact the most basic necessities food and water–sooner than we might imagine. The challenge is even greater for India given the stage of economic development we are in. The coming years will see a significant surge in our energy consumption most of which will be met by burning increasing quantities of fossil fuels.

If climate experts are to be believed, we are on the verge of losing all the economic gains made over the past couple of decades because India's economy ultimately sits on a bedrock of the subcontinent's ecology.

Environment and development are two sides of the same coin. The development of science and technology and ever increasing world population brought about tremendous changes in the earth environment. These changes upset the ecological laws, thereby disturbing the balance of human life. Any one of these cannot be sacrificed for the other. In the mad race for development, developed and developing nations alike are tremendously surging towards rapid industrialization at the cost of destroying the natural ecology of the environment. Since Development and environment pollution are inevitable and interdependent phenomenon, judicial attitude is geared around for drawing up of a line of control over both these contingencies.

Developmental attitudes causes increasing impact on the ecosystem, which is essential for well being of mankind. Stiglitz (2002) uses apt metaphor of dilapidated boats when he says "small developing countries are like small boats. Rapid capital market liberalization, in the manner pushed by the International Monetary Fund holes in their hulls have been repaired, before the captain has received training, before life vests have been put on board. Even in the best of circumstance, there was a high likelihood that they would be overturned when they were hit broadside by a big wave".

The current environmental crisis is world wide, among the most serious issues are the destruction of tropical rain forests, the depletion of the ozone layer, acid precipitation, soil erosion and depletion and global warming. In 1993, almost 1700 of the world's top scientists, including 104 Nobel Laureates, signed a "Scientists Warming to Humanity", which argued that current methods of production and consumption are unsustainable, that "human beings in the natural world are on a collision course' and that "a great change in our stewardship of the earth and the life on it is required, if vast human misery is to be avoided and our global home on this planet is not to be irretrievable mutilated".

Economic growth and development of a country or region may be realized through the optimum utilization of available resources and the state of technology, which in turn depends on the quality of institutional arrangement or structure that is being the force in production and distribution. Because there is a close connection between the institutions and technology employed and the efficiency of a market economy is directly shaped by the institutional framework. In the whole process of the economic progress, various types of institutions are to function and they work at various levels interdependently. It is evident that institutions are as important as technology for economic development and they evolve together interactively during the development process. The role of institutions in economic activity is that defining the parameters and processes through which agents use technology, natural and human resource are to increase the level of production and income. However, it should be noted that all the time all the institutions would not be equally effective.

Whenever some external substance interferes with the original one and deteriorates or destroys its quality then it is known as pollution. Environment pollution happens when external factors deteriorates the natural environment in any form. These human activities are like use of automobiles, combustion engines, industrial wastes, household wastes, fertilizers and others. When this environment pollution reaches to a great extent it takes the form of environmental crisis. It is at this stage, when a balance between biotic and abiotic substances gets imbalance and resulted into adverse effect on the lives of living being.

During the twentieth century, people have put a heavy burden on the earth's environment by giving priority to economic development, mass production, mass consumption and mass disposal. Population explosion, industrialization, unplanned urbanization, depletion of natural resources, deforestation, generation of sewages, liquid and solid waste materials, technological advancement etc are the major cause of poor quality of environment in metropolitan cities.

India is one of the fastest (3rd) growing economies of the world. It achieved a commendable growth rate (9%) in 2006 and the average growth rate in the Xth five year plan period (2002-03 to 2006-07) is about 7%. This is the highest rate achieved in any plan period. For the XIth plan the target is to achieve 9% growth rate. But there are several constraints in the way to sustainable growth.

'Industrial Revolution' gave mankind both luxuries and evils of push button living. Industries generate, use and discard toxic substances that lead to the generation of enormous wastes possessing hazardous characteristics scientifically proven to have adverse effect on human health and environment. The hazard industrial development and the anarchic operations of the world economy have had a disastrous impact an the global environment this has, also led to the generation of huge quantities of hazardous wastes that includes all kind of toxic materials such as pesticides, nuclear waste, bio medical waste etc. Development today is no longer only a comfort but also a liability. The reason: development brings in as many problems as it solves. The common scene across Asia today, to name a few consists of over flowing streets, lack of integrated planning.

For about Six decades, since independence, we have adopted a growth strategy, aiming at achieving the highest possible growth rate through exploitation of environment. Although the heavy heavy and basic industries were located mainly with a view to reducing regional disparities rather than increasing environmental pollution.

MODEL OF GROWTH

The old Paradigm of directed and regulated economic growth was replaced by a new paradigm of market-oriented growth, guided by profit motive, leading to flow of industries to areas where economies were available abundantly. The objective underlying this strategy is to maximize growth rate caring little for preservation of natural resources and bio-diversity. Private sector thus never calculates social costs.

Table 1
Impact of Development Activities on Environment

Development activities	*Impact on Environment*
Forest clearing and land selltement	Extinction of rare species of flora and fauna, create breeding grounds for mosquito.
Shifting cultivation in upland agriculture	Soil erosion in upland areas. Soil fertility decline. If shorter cultivation cycle, flooding of low land areas.
Agro-industries e.g. Sugar and tapioca	Air pollution due to burning of bio gas as fuel in mills, large amount of highly polluting organic with surface water pollution.
Use of pesticides	Organisms develop resistance & new control methods needed. Leads to complex and widespread environment problems. Biological magnification, poisoning.
Timber Extraction	Degrades land, destroys surface soil, reduces production potential of future forest.
Urbanization and industrialization	Concentration of population, making huge dent on rural environment, air and water pollution.
Water resources projects	Human settlement and resettlement spread of e.g. extensive irrigation, waterborne disease reduction of fisheries, physical changes, e.g. temperature, humidity.

Thus in our both models of growth, one of the direct growth and the other of market oriented growth, the economists did not assign any place to natural resource conservation and pollution abatement. We have to ensure higher growth rate without increase in environmental pollution. For this purpose, we have to change our approach as well as strategies in regard to higher economic growth and better environment, in such a way that one reinforces the other.

ENVIRONMENTAL PROBLEMS IN DEVELOPING COUNTRIES

The rapid industrialization in developmening countries results in irreparable contamination of surface, ground water, air and soil, the depletion of forests the discharge of hazardous effluents from industries, the uncontrolled dumping of contaminated solid waste, over-exploitation of natural resources, increasing number of unknown diseases caused by pollution, and new phenomenon like acid rain and so on. The situation existing in many developing countries is the complex growth of environmental pressures is the rapid population growth, unprecedented rates of urbanization and the urgent need to pursue economic development at whatever cost.

Proper management of the natural resources base is especially important in developing countries, because these countries can least afford the opportunity costs of irreversible loss of their renewable natural resources, or efforts to remedy environmental damage. But despite their increasing attention to environmental management many developing countries still cannot deal effectively with serious environmental problems they face. Important among these are the need to conserve and manage renewable natural resources (including forests, crop lands, grazing land, marine and fresh water fisheries and wild life) control of industrial and urban air, water and land pollution, public health aspects of developmental projects, occupational health and safety, involuntary resettlement of people living in project areas, the well being of tribal peoples and the preservation or salvage of unique cultural properties such as archaeological sites. Many environment losses are either completely irreversible or reversible only over a very long time. Among the more urgent environmental problems of developing countries are those relating to land and water resource. Other serious environmental problems also contribute to problem of human health and well being in developing countries. The costs of add on pollution control technologies are usually prohibitively expensive and do not resolve the basic deficiencies in the industrial processes, moreover such facilities are most often operated very inefficiently and with many health and safety hazards. The economies of many developing countries depend heavily on their natural resources base, especially in agriculture, forestry and fisheries.

Urban India began to colonise India, thousands of dams were constructed without once pausing to establish whether or not the nation was profiting from them, mines were dug indiscriminately pesticides such as DDT were sprayed in such massive doses that instead of insects, human began to die from the poisons. Nuclear reactors were built without worrying about the fact that there was no place to dump the lethal waste they would continually produce. As for our forests, almost anyone with an axe, bulldozer or servants at their disposal began to cut unprotected timber for sale to the highest urban bidder.

Similarly the slums are the result of urbanization. The urban population has grown 5 fold from 1951-2001. People come to urban area in search of employment. Urban areas are reeling under increasing pollution, traffic congestion making urban life unbearable. On going fast-paced urbanization under the influence of increasing globalization especially in developing countries of Asia is like a twin edged sword with multiple risks, and consequences of several kinds.

Slums by definition, are heavily congested residential settlements in which one find bad housing and poor family life, over crowding, crimes, insanitary conditions, begging, pollution, drug abuse, human abuse, lack of basic service, endangering public health, safety and morals. The steep rise in slum growth has led to grave threats mainly to the environment, law and order and to public health. As a consequence of environmental crisis, problems like excessive heat, excessive humidity, adverse effect on physical and mental development of human beings, danger to health of living beings. Extensive industrialization has resulted into polluted world. Three Indian cities are amongst the 15 most polluted cities of the world.

Urbanization promotes the influx of population from rural areas to urban centers because of push and pull factors operative in the respective area. This has resulted into urban poverty, fast growth of informal sector, and acute pressure on basic services which ultimately deteriorates urban environment. But we must not forget that the world's future is largely an urban future. Our ability to incorporate the urban poor in the development processes and assist in their growths will be highly significant if we are to achieve sustainable development.

Development comes through industrialization which in turn the main factor behind the degradation of environment. To resolve the issue, the experts worldwide have come up with a doctrine called 'Sustainable Development' i.e there must be balance between development and environment. The doctrine of 'Sustainable development' resolves the issue of such development at the cost of the environment by adopting an anticipatory measure to tackle the future hazards. Development and Environment must co-exist and go hand in hand. Economic growth is needed to realize development; it should be of Sustainable growth mode so that it does not result in the depletion of capital or environmental assets beyond the rate of regeneration of these assets. Not surprisingly, galloping soil sterility has overtaken millions of hectares of once fertile farmlands. And there is not one river in India whose water we can safely drink. For millions, landslides, floods and droughts make a mockery of the delusive projections of economists and politicians.

Sustainable development in a global economy requires a delicate long-term balance between human activity and nature's ability to renew itself. Future prosperity depends both on preserving ecological treasures for the benefit of future generations and on continued economic growth and innovation. For smooth running of society several legislations are passed. The law relating to environment has seen a paradigm shift over the concept of right to environment in India since 1980s; primarily after a series of global co-operative initiatives. The nation has not only enacted specific legislation on environmental protection, but has also created a new fundamental right to clean environment in the constitution.

The environment (protection) act 1986 provides for single focus in the country for the protection of environment and aims at plugging the loopholes in the existing legislation. In the developing countries like India, exploitation of natural resources is essential for economic progress for development. Right to wholesome environment is also a fundamental right under article 21 of the Indian Constitution.

To achieve the equilibrium between both, Hon'ble Supreme Court in series of landmark judgment has evolved some fundamental principles. The important principles evolved by the highest court of the country focuses on concept of:

- 'Sustainable development'
- 'Polluter pay'
- 'Precautionary'
- 'Strict liability'

With growing environment concerns in India policy makers have evolved the system of Environment Impact Assessment which seeks to bring about a balance between industrialization and environmental conservation. As hazardous industries become a characteristic of 21st Century world, and India is in the race of rapid industrialization which is accompanied with massive environment and social burdens, monitoring tools like Environment Impact Assessment (EIA), assumes great significance in ensuring sound economic development without compromising on environmental and social costs.

Some significant features of the Act are:

- Making top Management responsible for environment protection and liable for prosecution.
- Providing stringent punishment and penalties.
- Empowering enforcement agencies to effect closure of polluting industries.

NEW POLICY FOR DISCOURAGING ACQUISITION OF FARMLAND FOR SEZS

In the backdrop of controversies over setting up of SEZs, the Central Government unveiled on November 25 2007, a new policy which provides for discouraging acquisition of agricultural land for non-farming purposes and setting up a commission to look into resettlement aspects. The National Rehabilitation and Resettlement Policy 2007 also entails setting up of a committee to review and monitor the progress of implementation and resettlement schemes across the country.

"Acquisition of agricultural land for non-agricultural use in a project may be kept to the minimum 3 multi-cropped land. Land may be avoided to the extent possible for such purposes, and acquisition of irrigated land, if unavoidable, may be kept of the minimum". The policy states. "Waste land, degraded land or un-irrigated land should be the option for acquisition." This assumed significance as acquisition of farming land for SEZs has triggered violent protests all over India. The new policy envisages setting up a Rehabilitation Commission at the national level to look into the Settlement issues of the families displaced from their land following its acquisition for projects.

Problems in Implementing the Policies

The Environment Protection Act, 1986 has always remained a paper tiger because of enforcement authorities in India could not effectively implement them it can be a powerful tool but has many loop holes and weaknesses. Law is a regulator of human conduct but no law can indeed effectively work unless there is an element of acceptance by people in society. There are certain problems such as:

- Deficiency of environmental education knowledge among enforcing authorities.
- Lack of responsibility due to ecological illiteracy on part of community.
- Lack of provision for training of professional manpower in the field of environmental education.
- Lack of coordination among the administrative institutions and educational institutions on environmental activities.
- Shortage of human and financial resources.
- Improper technology
- Lack of overall proper planning

ECO-FRIENDLY WORK SOLUTIONS

Developing Countries are in need of 'home grown' solutions which are in harmony to live well can not achieve high standards of living at the expense of environmental and ecological integrity. The situation is alarming and requires immediate action, legislation, execution and planning so that we may justify ourselves.

Therefore the need of a comprehensive approach to the problem of environmental protection is required. They can be

- Innovative strategies to address the problems are necessary.
- People's participation is very helpful and urgent need here (NGOs, SHGs).
- Development of minor irrigation and utilization of created irrigation potential.
- Rainwater harvesting and conservation for the development of rainfed areas water shed approach.
- Reclamation or development of problem soils/lands.
- Encouragement to the private sector for effective extension services.
- Encouragement to organic farming.

- Detailed indigenous knowledge and greater skills in blending modern and traditional technologies to enhance productivity efficiency of farming community.
- Encouraging sanitary landfills, biogas generating technologies and vermicomposting.
- Offering good incentives for the local bodies and the government, to drive benefit by utilizing the wastes.
- There are two major areas on which research and development efforts are needed.
  - (i) Genetic resource management and bio-technology.
  - (ii) Natural resource management and agro ecology.
- The local body with the help of Govt. should chalk out the plan to divert and dispose of sewage.
- Proper drainage system be developed to get rid of water logging problem.
- Help more in pollution prevention than pollution control.
- Use eco friendly products.
- Save electricity by not wasting it when required because, electricity saved is electricity generated without polluting the environment.
- Promote reuse and recycling wherever possible and reduce the production of wastes.
- Use rechargeable batteries. Rechargeable batteries will reduce metal pollution.
- Environment education should result in the development of an ecological ethics a change in attitude of man towards man, society and nature, in realization of man as a part of nature, not alien to it.
- Adequate funds should be provided by the UGC, Dept. of science and Technology Central and State Governments for effective implementation of environmental education programme.

- Developing forest cover in the urban areas will result in reducing the air and sound pollution in cities and people living there will have benefits like purification of air, recreation etc available locally.
- The existing regulations and laws to protect forest and wild life should be implemented by the concerned government agencies vigorously and in spirit.
- People dependent on the forest should be trained to use the resources in self sustainable and viable manner.
- Provide green cover along the main roads, railway tracks and so on.
- Provide thick foliage between road dividers.
- The public carriers like buses and autorickshaw's should compulsorily use fuels like CNG in place of petrol and diesel so as to reduce the pollution.
- Nuclear power plants should be developed to generate power on large scale.
- The Government should install equipment to measure the pollution and presence of particles and harmful gases in the atmosphere. So that the corrective actions are taken on a timely basis.
- Legislation, though in place, should be implemented in letter and spirit for disposal of effluent by the industries. Pollution industries should be controlled in the vicinity of the large water bodies.
- An in-depth study of impact of large dams and reservoirs on the ecology should be studied before construction.
- The mass means of transportation should be developed like metro railways, buses etc in order to reduce the traffic and use of private vehicles in the cities. This will drastically reduce the pollution caused by vehicles.
- Use of windmill to generate power.

- Use of solar energy in lighting, heating and running small domestic appliances.
- Use of biogas in households.
- Use of nuclear power to generate electricity in place of coal, naphtha or lignite.
- Teachers, parents, doctors, engineers, planners, administrators and scientist and general public should be involved in environmental education programme.
- The interdisciplinary research projects on environmental problems should be undertaken.
- The Govt. should establish environmental research centers in every state. A data bank and eco systems information base/system should be developed, so that the required information is available at a central place.
- Seminars, conferences and workshops should be organized from time to time on environmental education.
- The general public should be educated about the environmental degradation and improvement of the environmental quality through mass media by incorporating them through regular programme of AIR, TV and visual aids etc, documentary films should be prepared on these themes and exhibited for the benefit of the people.

CONCLUSION

Development and Environment are interrelated to each other, as one hand development affects the environment and on the other, more stress on environment protection would close the door to development. The two are so significant that one of them cannot be compromised for the growth of the other. To achieve the objective of sustainable development lies with technological know-how and social commitment of research institutes, environmental regulating agencies, as well as the government's determinations in devising scientifically sound, pro-active or preventive strategies and policies for sustainable development. Sustainable rate of growth is that rate of growth which can be continued over a long period of time, for making growth sustainable in India, we have to make growth inclusive by tackling the problems of poverty, inequality and improving the environment. Proper planning, implementation, management of policies and public participation can ensure a sustainable system.

In matter of ecology and environment and considering the principle of sustainable development, no person or organization, however, high and mighty they man be cannot be permitted to flout the law of the land. The variations in the environment include the following aspects like urban and rural industrial and the agricultural, the slum dwellers and the sophisticated. Education at different levels should take cognizance of these factors and if they are ignored, it will not be meaningful. Education programmes and courses should be planned and implemented, taking into consideration, the very vital role played by the environment in the life of learner.

The United Nations Conference on environment and development held in Rio de Janeiro in 1992 and popularly known as 'Earth summit' followed by the world summit on sustainable development at Johannesburg in 2002, just 10 years after the first summit, have highlighted the key issue of global environmental concern and have attracted the attention of the general public towards the deteriorating environment. Any government at its own level cannot achieve the goals of sustainable development until the public has a participatory role in it. Public participation is possible only when the public is aware about the ecological and environmental issue. A drive by the government to ban the littering of polythene cannot be successful until the public understands the environmental implications of the same. The public has to be educated about the fact that if we are degrading our environment we are actually harming our ownselves.

This is because we are a part of the complex network of environment where every component is linked to one another. It is all the more important to educate the people that sometimes the adverse impact of environment are not experienced or noticed until a threshold is crossed. There is a Chinese proverb. "If you plan for one year, plant rice, if you plan for 10 years, plant trees and if your plan for 100 years, educate people".

In the present era, the human race is faced with the danger of environmental hazard. This has belittled the freedom to live in clean environment. We are to balance or coordinate properly between environment and development should be synonymous to each other when we talk about human being, we will have to include ecosystem.

The society shall have to prosper, but not at the cost of the environment and in the similar vein, the environment shall have to be protected but not at the cost of development of the society. It is important to strike a balance between the two i.e. development on one side and pollution free environment on the other.

REFERENCES

Adam D. (2000), Planning Between Cultural Paradigms: Traditional Knowledge and Transition to Ecological Sustainability, Dissertation Abstract International. 60(7): January.

B. N. Neelima, Communication for Environmental Education, JRF, Deptt. of Journalism, Sri Padmavathi Mahila University, Tirupathi.

B. R. Kukreti (1993), Environmental Education – A Blue Print, University News, August 16.

Bharat B. Dhar (2000), Environmental Education in India, University News, 38(6): February.

Bhatia A. (2000), Women's Development and NGO's, Rawat Publications, Jaipur and New Delhi.

Francois N. (2000), Cultivating the Human Environmental for the Adoption and Use of Information Technology, *Dissertation Abstract International*, 60(7): January.

G. R. Chhatwal (1998), Encyclopaedia of Environmental Education, Anmol Publications Pvt. Ltd. New Delhi.

Iyer S. (1996), Blue Print for Environmental Action, Social Welfare 40(8): June.

Jain R. (1997), Sustainable Development, Policy and Practices 1st Ed. New Age International Publishers, Australia.

K. B. Powar (1993), Environmental Issues and University, University News, November 22.

Onkar N. Srivastava (1992), Environment and Sustainable Development, University News, January 13.

Prakash J., and Srivastava S. K. (1999), Environmental Problems, Policies and Strategies, Mohit Publications. New Delhi.

Rs. 60 Crore for Environmental Studies (1990), University News, November 12.

S. Lakshmi (1991), Challenges in Indian Education.

Sharma S. (2001), Managing Environment: A Critique of "The Tragedy of Commons", *Journal of Human Ecology*, 12(1).

Subramanyam V., and K. R. Rama Moham (2001), Tribal Ecology and Food Scarcity, *Journal of Human Ecology*, 12(5): 351-356.

Vishwanandha K. R. (2000), Environmental Impact on the Economic Life of Yeruka Tribe. *Journal of Human Ecology*, 12(4): 267-270.

Yogendra K. Sharma (2001), History of Problems of Education (Volume 2), Kanishka Publication.

Ali Sabir, and S. N. Singh (1998), Major Problems of Delhi Slums Uppal Publishing House, New Delhi.

D. Bhugra, and Anastasia Mastrogianni, Globalization and Mental Disorders, *British Journal of Psychiartry*, 184: 10-20.

Desh Bandhu, and N. L. Ramnathan (1982), Education for Environmental Planning and Conservation, Indian Environmental Society., New Delhi.

M. A. Farok (2005), Issue in Environmental Education, Arise Publishers and distributors, New Delhi.

Asthana D. K., and Asthana M. (2001), Environment Problems ad Solutions. S Chand and company, New Delhi.

Chhokar K. B., and Pandya N. (2004), Raghunathan, Understanding Environment, Sage Publication, New Delhi.

Deb S. C. (2003), Environmental Management Jaico Publishing House, Mumbai.

Ghanta R., and Rao D. B. (1998), Environmental Education Problems and Prospects, Discovery Publishing House, Delhi.

Ghosh A. (2000), Environment ad Development, A.P.H. Publishing Corporation, New Delhi.

Kaushik A., Kaushik C. P. (2004), Perspectives in Environmental Studies, New Age International Publications, New Delhi.

Singh P. P., and Sharma S. (2004), Teaching of Environment. Deep and Deep Publication, New Delhi.

Prabhakar V. K., (2002), Environmental Pollution, Ammol Publication, New Delhi.

Honley Nick, (2002), Robert Colin, Issues in Environmental Economics, Black Well Publishers, UK.

Bibliography

http://www.scacobgy;org/?gclid=(New 3 nig, 7ko8cfsuy about kxdrv)

http://www.gisdevelopment.net/application/environment/pp/envp 1001 .htm

http://www.unrptie.org/ozonactin/events/ozoneday/2006-events.htm

http://www.m-w.com/difinary/environment

http://www.infochangeindia.org/analysisl35.jsp

http://envfor.nic.in/report/report/html

http://www. inforchangeindia.org/environmentbp.jsp

http://www.indiabudget.nicin/es98-99/environ.htm

6

WHO NEEDS CONSERVATION AND FROM WHOM? OLIVE RIDLEY TURTLE CONSERVATION IN ORISSA SEACOAST

Pradip Swarnakar

Like other species *Homo sapiens* are not homogenous in terms of their behaviour. This is true for their attitude towards other species and fellow humans. For the last few decades *Homo sapiens* are more worried about their surrounding environment. Let's take a look at the possible viewpoints of humans towards other species.

1. Humans are considered fittest in the survival struggle. Throughout the history of mankind they have a contesting attitude towards other species. The idea behind this survival principle is "live and let them die" which always uses nature as resources. Nowadays humans are worried regarding environment because resources are decreasing which eventually endanger their existence.
2. Another group of human beings are concerned for the environment, triggered by the idea of "live and let live". This group of humans is saving the life of the endangered species for the sake of that species

which may not have an immediate benefit to humankind. This is the expression of coexistence with nature.

3. A group of humans have moved one step further from the above in terms of conservation practices. In this case, sometimes human beings are negatively suffered or affected by the forest/animal and other kind of conservation. It becomes more altruistic in the form of "die and let live". Obviously, human beings by its own nature are unwilling to contribute voluntarily to this kind of altruistic attitude. So, why this is happening? A group of human beings popularly known as policy makers or lawmakers (for the benefit of mankind and overall biosphere) impose certain law or policy to another group of individual. In this case two groups of humans put themselves in conflicting relationship over the conservation of other species.

Hence the questions arises "who needs conservation?" Is it really a conflict between two contesting species or a conflict within the heterogeneous groups of the same species? This paper explores the nature of inter and intra human conflict with respect to turtle conservation and raises the issue of conflicting interest within different groups of human beings.

OLIVE RIDLEY TURTLE

The olive ridley turtle (*Lepidochelys olivacea*) is considered the most abundant sea turtle in the world, with an estimated 800,000 nesting females annually[1]. The olive ridley may be the most abundant sea turtle on the planet, but some argue that it is also the most exploited. According to the Marine Turtle Specialist Group (MTSG) of the IUCN, there has been a 50 per cent reduction in their population since the 1960s. The olive ridley is listed as endangered by the World Conservation Union (IUCN)[2]. Olive ridleys often migrate great distances between feeding and breeding grounds. In India the mass nesting of olive turtle is called *arribadas*. The mass nesting site is located at three sites of Orissa coast – Gahirmatha (Kendrapada district), Rushikulya (Ganjam district) and Devi river mouth (Puri district) (Pandav *et al.*, 1998)[3]. Gahirmatha, used to be one of the largest arribada nesting sites in the world during January to May nesting season. In Gahirmatha, between 1996 and 2002, the average size of nesting females declined at that site, indicative of a declining population (Plotkin ed. 2007)[4]. In 1997, 1998 and 2002 the turtles skip the annual rituals and there was no mass nesting at all (The Hindu 2005)[5]. The principal cause of the historical, worldwide decline of the olive ridley sea turtle is long-term collection of eggs and killing of adults on nesting beaches. The recent failure of arribadas in 1997 and 1998 suggest a potential or imminent decline in the population, consistent with fishery-related mortality of at least 1,00,000 turtles since 1994, and 10,000-15,000 turtles per year since 1999 (Shanker *et al.*, 2004)[6]. Expansion of the shrimp trawling fishery in the eastern Indian Ocean in the mid-1970s has resulted in numerous olive ridley deaths. Over 10,000 olive ridley carcasses a year have been counted on the Orissa coast since 1999 (Wright and Mohanty 2006)[7]. Much of this mortality is attributed to drowning in trawl nets (Pandav and Choudhury, 1999)[8], and gill nets (Wright and Mohanty, 2002)[9]. Although the present death of the turtle is mainly incidental, before 1970s turtles were killed for their meat and eggs for local and Kolkata market (Kar, 2001)[10]. In Gahirmatha, hundreds of thousands of eggs were once collected annually and revenue (*andakara*) was paid to the local zamindari before Independence and to the Orissa Forest Department until seventies (The Hindu 2003)[11].

CONSERVATION EFFORTS

Olive ridley turtles are protected by various international treaties and agreements as well as national laws. The main legislations in operation in the state are the Indian Wild Life (Protection) Act, 1972 (WLPA), the Orissa Marine Fisheries (Regulation) Act, 1982 and the Orissa Marine Fisheries (Regulation) Rules, 1983. The WLPA declared the consumption, trade, hunting and injury of turtles as prohibited and after enforcement there was a sharp decline in turtle trade in Orissa by the 1980s (Kar 2001). In 1997, the Government of Orissa declared Gahirmatha and its waters as the Gahirmatha Marine Sanctuary (GMS) under section 26(1) (b) of the WLPA. The marine sanctuary was divided into a core area measuring 725.5 km^2 and a buffer zone measuring 709.5 km^2. The total area of the GMS measures 1435 km^2 (Sridhar *et al.* 2005)[12]. The core zone of the sanctuary extending 10 km from the shore (a no-fishing area) and an additional 10 km was declared a buffer zone. In the mid 1990s the Government of USA modified Section 609 of Public Law 101-162 of the Endangered Species Act, 1973 and made it mandatory for countries exporting shrimp to the U.S. (Bache and Frazier 2006)[13]. This includes the usage of turtle excluding devices (TED) in the fishing net.

INTERNATIONAL AGENCY: GREENPEACE

In recent past international environmental organization Greenpeace is campaigning over the issue of turtle conservation in Orissa coast (Greenpeace News 2004)[14]. With the help of local youths their ocean campaigners are building awareness among the people. At the time of nesting they are actively protecting the eggs of the turtle. However, their effort is not sufficient to cover the long beach area. The most important thing is that they are more concerned regarding conservation of turtle. They criticize government policies and their enforcement (Greenpeace News 2006, 2007, 2008)[15] and industrial intervention (Greenpeace News 2007)[16] which endanger turtles' life. However, in their past campaign and subsequent newsletter publications they never mentioned the problem of the local fishermen who are affected negatively from the conservation policies and practices.

LOCAL COMMUNITY: FISHERMEN

The primary livelihood of the most of sea coast inhabitants is fishing. Within the fishermen community they have large, medium and small fishermen. Due to the WLPA livelihood of the fishermen community particularly of the small and marginal farmers are negatively affected. Although they have no intension to kill or catch turtle, incidental death is occurring due to fishing-nets and turtles are dying as a collateral damage. Moreover, WLPA does not make a clear distinction between incidental or accidental capture in fishing nets and poaching. Therefore, it is the right of surveillance agency to penalize the person as poachers if they found turtle in their fishing nets, irrespective of whether the catch takes place within or outside a protected area. Furthermore, enforcing officers empowered by WLPA can harass fisher folk and confiscate their fishing vehicles. In addition to that using TED in their fishing net also reduced the fishing catch (Shanker and Kutty 2005)[17].

Due to this law, fishermen communities are suffering in two ways. First, with the help of small capacity boat they are unable to cross a long reserved sanctuary (core area) boundary to go to government specified fishing area. Second, if they fall under law keepers' gaze they are punished heavily. The forest department official most of the time confiscate their boat and ask for hefty amount of money to release them along with fear of imprisonment and fines. This situation jeopardizes their traditional livelihood and it is said that this ban affect more than 20,000 fishermen. Consequently, earning male member, who are unable to struggle this battle are migrating to other places and in extreme cases committing suicide (Down to Earth 2007)[18].

OTHER AGENCIES

Industrialists and investors are communities within *Homo sapiens,* who are responsible for economic development of the nation. Let's have a look at their activities in relation with turtle conservation. The 'Wildlife Conservation Strategy' adopted by the Government of India in 2002 states that lands falling within 10 km of any protected area should be declared an ecologically fragile zone. Such zones would have the possibility of regulating environmentally destructive activities while permitting benign ones. There are also laws for the protection of the coastal environment such as the Coastal Regulation Zone Notification, 1991 and the Water (Prevention and Control of Pollution) Act, 1974. These regulate the activities of industries in the coastal zone (Sridhar *et al.*, 2005).

Near Gahirmatha marine sanctuary TATA has proposed to set up Dhamra port. It is interesting to note that the proposed projects is located twelve kilometres north of the nesting beach at Gahirmatha, and it lies just outside the boundary of the sanctuary. The project was granted environmental clearance not by the Ministry of Environment and Forests (MoEF), but also by the Ministry of Surface Transport (MoST) of the Government of India. At present, the Indian corporate bodies involved in the port project at Dhamra are believed to include ICICI Bank, construction major Larsen & Toubro (L&T) and Tata Steel (Sekhsaria 2005)[19].

Another of India's large industrial houses, the Reliance Group has planned to explore oil and natural gas. The company will carry out exploratory drilling, one at NEC-25 which is 75 km off the Gahirmatha Sanctuary, and another at MD-10, which is around 60 km off the Rushikulya river mouth (Sekhsaria 2004)[20]. The RIL officials advocate that companies such as Oil and Natural Gas Corporation and Oil India drilled many exploratory wells since 1976 within the close proximity of Gahirmatha Sanctuary and there was no evidence of any adverse effect on the turtles (The Hindu 2004)[21]. However, the nature conservationists like Greenpeace are claming that these industrial interventions will harm the mass nesting of olive turtle and endanger their existence.

In addition to Reliance and Tata groups other industries are also responsible for endangering turtle's life. Some examples include the reported release of untreated effluents by Oswal Chemicals and Fertilisers Ltd at Paradip (Greenpeace News 2000)[22] and of Jayshree Chemicals at Ganjam (The Hindu 2008)[23]. All these activities not only endanger the turtles' life but also negatively affect fishermen livelihood as most of these industrial activities endanger the life of flora and fauna of marine ecology.

WHO NEEDS CONSERVATION?

Most part of the world have been modified, managed and in some instances, improved by humans for centuries. In this process other species have been either extinct or their life have been threatened. The noble ideology behind the philosophy of conservation emerged from the idea of "live and let live". However, the equation is not so simple. In the present situation turtle and human beings are not struggling with each other for their existence. On the contrary, various social groups, sometimes with conflicting interest, play important role over the issue of turtle conservation. These groups include fishermen, international conservationist organization Greenpeace, industrial groups like TATA and Reliance, and state government.

It was believed that local fishermen along with other interest groups were involved in the selling business of turtle meat and egg before 1970's. Their motivation comes from "live and let die" philosophy. After the introduction of wildlife protection laws (which promote turtle conservation) direct killing of turtle has been stopped, but the incidental death is still taking place. However, if turtles are found in their fishing net, the wildlife protection laws and its implementing agencies do not distinguish poacher and legal fisherfolk. It not only threatens the livelihood of fisherfolk but also demand an altruistic sacrifice from the fishermen community "to die and let turtle live". In addition to this, the industrial activities of TATA, Reliance and other groups endanger both the life of fishermen and turtle. Interestingly government has double standard towards two different social groups: fishermen and industry. On the one hand it is encouraging industrial groups and on the other it restricts fishermen from their livelihood. Moreover, international agency (Greenpeace) is only concerned about endangered turtle not to any other vulnerable human groups. In the present situation it is not easy to solve the puzzle. So, looking at the future sustainability of human and turtle a more comprehensive policy, involving all stakeholders, should be taken. Only then the idea of "live and let live" will be meaningful for both humans and turtle.

NOTES & REFERENCES

1. See NOAA Fisheries (2008), Office of Protected Resources, *http://www.nmfs.noaa.gov/pr/species/turtles/oliveridley.htm*, Downloaded on 14 May 2008.

2. IUCN (2007), *2007 IUCN Red List of Threatened Species.* <*www.iucnredlist.org*>. Downloaded on 14 May 2008.
3. Pandav B., B. C. Choudhury, and K. Shanker (1998), 'The Olive Ridley Sea Turtle (*Lepidochelys Olivacea*) in Orissa: An Urgent Call for a Intensive and Integrated Conservation Programme', *Current Science*, 75: 1323-1328.
4. Plotkin P. T. (2007), *Biology and Conservation of Ridley Sea Turtles*, Baltimore, MD: The Johns Hopkins University Press.
5. Basu Somika (2005), Fettered Flippers Along the Orissa Coast, *The Hindu*, July 10.
6. Shanker K., B. Pandav, and B. C. Choudhury (2004), 'An Assessment of Olive Ridley (*Lepidochelys Olivacea*) Nesting Population in Orissa, India', *Biological Conservation*, 115: 149-160.
7. Wright B., and B. Mohanty (2006), 'Operation Kachhapa: An NGO Initiative for Sea Turtle Conservation in Orissa', In. Shanker K., and Choudhury B. C., (Eds.), *Sea Turtles of the Indian Subcontinent*, Hyderabad, India: Universities Press, pp. 290-302.
8. Pandav B., and B. C. Choudhury (1999), 'An Update on the Mortality of the Olive Ridley Sea Turtles in Orissa, India', *Marine Turtle Newsletter*, 83: 10-12.
9. Wright B., and B. Mohanty (2002), 'Olive Ridley Mortality in Gill Nets in Orissa', *Kachhapa*, 6: 18.
10. Kar C. S. (2001), Review of Threats to Sea Turtles in Orissa. In: *Proceedings of the Workshop for the Development of a National Sea Turtle Conservation Action Plan, Bhubaneswar, Orissa* (Eds. K. Shanker & B. C. Choudhury), pp 15-19. Wildlife Institute of India, Dehradun.
11. Shanker Kartik (2003), What ails the Ridley?, *The Hindu*, June 8.
12. Sridhar A., Tripathy B., and Shanker K. (2005), 'A Review of Legislation and Conservation Measures for Sea Turtles in Orissa , India', *Indian Ocean Turtle Newsletter*, 1: 1-7.
13. Bache S. J., and J. Frazier (2006), 'International Instruments and Marine Turtle Conservation', In: *Sea Turtles of the Indian Subcontinent* (Eds. K. Shanker & B. C. Choudhury) Universities Press, Hyderabad. India.
14. Greenpeace News (2004), 'Orissa-The Mass-Nesting Destination of Olive Ridley Turtles'.

15. Greenpeace News (2006), 'Turtle Massacre in Orissa: Greenpeace Hold Chief Minister Responsible'.

15. Greenpeace News (2007), '11.21 Crores to End Turtle-Human Conflict: Greenpeace Demands Budget Allocation', March 7.

15. Greenpeace News (2008), 'Turtle Mortalities Skyrocket as Government Sits Idle', March 19.

16. Greenpeace News (2007), 'TATA's Dhamra Port EIA Seriously Flawed, Project Must Not Go Ahead: Greenpeace', May 31.

16. Greenpeace News (2007), 'Greenpeace Ask TATAs to "Take a Turtle-Friendly Look at Life"', June 19.

17. Shanker K., and Kutty R. (2005). 'Sailing the Flagship Fantastic: Myth and Reality of Sea Turtle Conservation in India', *Maritime Studies*, 3(2) and 4(1): 213-240.

18. Mackinnon Benjamin (2007), 'Olive Ridley Preservation Hits Fishermen in Orissa', *Down to Earth*, January 1.

19. Sekhsaria P. (2005), 'The Dhamra Port in Orissa', *Indian Ocean Turtle Newsletter*, 1: 14-16.

20. Sekhsaria P. (2004), 'Reliance vs the Olive Ridley Turtle', *InfoChange News & Features*, June 14.

21. Das Prafulla (2004), 'Oil Exploration Near Olive Ridley Turtle Nesting Sites Causes Concern', March 20.

22. Greenpeace News (2000), 'Bahut Ho Gaya!', April 1.

23. The Hindu (2008), 'Preparing for Hatching of Olive Ridley Eggs', April 18.

7

A STUDY ON ENVIRONMENTAL DEGRADATION IN BARHUR SIPCOT INDUSTRIAL REGION

I. Sundar & R. Mohanraj

INTRODUCTION

Environmental degradation expressed as the pollution of the natural resources and eroding the capacity of the eco-system support healthy living in the wake of the proliferation of industries in residential neighborhood is a significant indicator of the suffering that industrialization imposes on generations of people to come. Industrialization associated with technological progress is itself a major cause of environmental pollution even as technological progress and innovation could also be part of the solution. Usually development of industries leads to both positive effects as well as negative effects. The positive effects may be in the form of employment generation, increase in income in the hands of local people, activity diversification, business diversification, increase in overall condition of living, development of subsidiary occupation and so on. Hence, there is a need to analyze some positive effects of establishment of SIPCOT industrial complex in Dharmapuri region for the purpose of present study. This study aims at analyzing the status of land degradation in the study villages inconsequence of industrial pollution.

It is significant to note that the study of socio-economic impact of industrialization is a subject of frequent discussion today. The participants are being planners, policy makers and politicians. This study analyses the nature and effect of industrial pollution on villages situated around the Bargur SIPCOT industrial region.

Objectives

1. To examine the households' views on positive and negative externalities of SIPCOT industrial complex on local area development.
2. To study the households' perception on land degradation, water pollution and air pollution in consequence of establishment of SIPCOT industrial complex.

Methods and Materials

The researcher has selected Bargur SIPCOT complex for the purpose of the present study. In order to assess the impact of SIPCOT industrial complex on local residents, six villages are selected which are located within a distance of 2 km radius from the SIPCOT industrial complex. Totally 300 households are selected from the 6 villages. The relevant data have been collected from the respondents by employing a well-structured interview schedule. The collected data have been classified and tabulated with the help of computer programming.

RESULTS AND DISCUSSION

A study of data in Table 1 indicates the Village wise respondents' views on negative externality of SIPCOT Industries complex on local area development. In general, industrial development leads to positive development as well as negative externalities. It is observed that out of the total 300 households, 28.67 per cent of them feel that the negative externality of establishing SIPCOT industrial region in Bargur in the form of loss of traditional land and household displacement.

Table 1
Village Wise Respondents' Views on Negative Externality of Starting SIPCOT Industrial Complex on Local Area Development

Village name	*Loss of traditional land and displacement*	*Property damage and household behavioural change*	*Conflict between local residents and pollution control board*	*Conflict between local residents and industries management*	*Total*
Jagadevi	18 (36.00)	13 (26.00)	9 (18.00)	10 (20.00)	50
Nagamangalam	13 (26)	12 (24.00)	10 (20.00)	15 (30.00)	50
Malapatti	15 (30.00)	8 (16.00)	9 (18.00)	18 (36.00)	50
Gandhi kuppam	13 (26.00)	12 (24.00)	14 (28.00)	11 (22.00)	50
Duvara palli	12 (24)	17 (34)	13 (26.00)	8 (16.00)	50
Venkatapuram	15 (30)	23 (46)	8 (16)	4 (8)	50
Total	86 (28.67)	85 (28.33)	63 (21)	66 (22)	300

Source Computed: Figures in parentheses denote row percentage

The State Industrial Promotion Corporation of Tamilnadu acquires their lands so they are displaced from their land. It is significant to note that 28.33 per cent of the households realize the negative externality of establishing SIPCOT industrial region in Bargur in the form of damage to their property and development of behavioural changes in consequence of living in the polluted environment. It is remarkable to observe that 21 per cent of the households realize the negative externality of starting SIPCOT industrial region in Bargur in the form of development of conflict between local residents and pollution control board. The pollution control board does not properly monitor the activities of industries towards releasing pollutants; so the people of local area make frequent compliant to the pollution control board. Moreover, 22 per cent of the households observe the negative externality of starting industries in SIPCOT region in the form of development of conflict between local residents and industrial management. This happens when industries are frequently releasing toxic pollutants into the water bodies, land and releasing toxic gases into the atmosphere. As a result, local people make conflict with industrial management to reduce the level of releasing pollutants. Majority of the households of Venkatapuram village observe the negative externality of starting SIPCOT industrial region in Bargur in the form of damage to their property and development of household behavioural change. Majority of the households of Malapatti village and Nagamangalam village visualize the externality of starting SIPCOT industrial region in Bargur in the form of development of conflict between local residents and industrial management along with loss of the traditional land and displacement.

A study on data in Table 2 indicates the village wise respondents' views on positive externality of starting SIPCOT industrial complex on local area development. It could be noted that out of the total 300 households, 23.67 per cent of them observe the development of SIPCOT industrial complex in the form of local employment generation along with

Table 2
Village Wise Respondents' Views on Positive Externality of Starting SIPCOT Industrial Complex on Local Area Development

Village name	*Increase in employment and development of non-farm activities*	*Business development and development of institutions*	*Increase in living condition*	*Increase in property value*	*Development of infrastructure facilities*	*Total*
Jagadevi	20	7	9	8	6	50
	(40)	(14)	(18)	(16)	(12)	
Nagamangalam	8	17	6	7	12	50
	(16)	(34)	(12)	(14)	(24)	
Malapatti	11	8	5	9	17	50
	(22)	(16)	(10)	(18)	(34)	
Gandhi Kuppam	16	6	9	12	7	50
	(32)	(12)	(18)	(24)	(14)	
Duvara palli	7	8	12	10	13	50
	(14)	(16)	(24)	(20)	(26)	
Venkatapuram	9	8	12	14	7	50
	(18)	(16)	(24)	(28)	(14)	
Total	71	54	53	60	62	300
	(23.67)	(18)	(17.67)	(20)	(20.67)	

Source Computed: Figures in parentheses denote row percentage

development of non farm activities and 18 per cent of the households realize the positive externality of development of SIPCOT industrial complex in the form of business development along with development of health and educational institutions. It could be noted that 17.67 per cent of them state that positive of them state that positive externality of development of SIPCOT industrial complex in the form of an increase in the condition of living among the local households and 20 per cent of them state it as an increase in property value. Moreover, 20.67 per cent of them observe the positive externality of SIPCOT industrial development in the form of development of local infrastructure facilities like roads, electricity, drinking water, institutions and so on. A considerable number of households of Jagadevi village (40%) and Gandhikuppam village (32%) observe the positive externality of SIPCOT industrial development in the form of increase in local employment along with development of non farm activities. A considerable number of households of Malapatti village observe the positive externality of SIPCOT industrial development in the form of development of local infrastructure.

Data presented in Table 3 indicate the village wise respondents' views on land degradation. It could be noted that out of the 300 respondents, 13.67 per cent of them hold the view that land degradation leads to loss of soil fertility. Of the total 300 respondents, 15.00 per cent of them hold the view that land degradation results in ground water pollution and 15.67 per cent of them assert that land degradation creates acidity in soil and its consequence results in low water retention capacity.

Of the total 300 respondents, 41.00 per cent of them realize the land degradation in terms of change in the structure of soil and low crop productivity. Moreover, 14.67 per cent of the respondents realize the land degradation in terms of accumulation of organic and inorganic solid waste. Majority of the households of Jagadevi village (42%), Nagamangalam village (48%), Gandhikuppam village (44%) and Venkatapuram village (42%) observe the land degradation in the form of changes in the structure of soil and low crop yield.

Table 3
Village Wise Respondents' Views on Pattern of Land Degradation

Village	*Loss of soil fertility*	*Ground water pollution*	*Acidity and low water retention capacty*	*Changes in structure of soil and low crop yield*	*Accumulation of organic and inorganic solid waste*	*Total*
Jagadevi	6 (12)	8 (16)	8 (16)	21 (42)	7 (14)	50
Nagamangalam	5 (10)	7 (14)	9 (18)	24 (48)	5 (10)	50
Malapatti	6 (12)	9 (18)	10 (20)	17 (34)	8 (16)	50
Gandhikuppam	8 (16)	6 (12)	8 (16)	22 (44)	6 (12)	50
Duvarapalli	10 (20)	7 (14)	5 (10)	18 (36)	10 (20)	50
Venkatapuram	6 (12)	8 (16)	7 (14)	21 (42)	8 (16)	50
Total	41 (13.67)	45 (15.00)	47 (15.67)	123 (41.00)	44 (14.67)	300

Source Computed: Figures in parentheses denote row percentage.

A keen observation of data Table 4 indicates the village wise respondents' views on the effect of land degradation on living beings. It could be noted that out of the total 300 respondents, 17.33 per cent of them realize the problem of land degradation in terms of bad odour and vomiting sensation and 22.00 per cent of them suffer due of skin diseases. The problem of skin disease affects about one fourth of the respondents of Venkatapuram village (26.00%). Of the total 300 respondents, 25.67 per cent of them realize the problem of land degradation in terms of development of health hazards to the livestock. Moreover, 35.00 per cent of the respondents refer to the problems of land degradation in terms of parasitic disease infection, due accumulation of polluted water on land along with other industrial waste. Majority of the households of Jagadevi village (40.00%) and Nagamangalam village (48.00%) argue that the problem of land degradation in terms of parasitic disease infection due to accumulation of polluted water which are the breeding places for various pathogens.

Data presented in Table 5 indicate the village wise respondents views on alternative ways of preventing land degradation. It could be noted that out of the total 300 respondents, 18.33 per cent of them hold the view that land degradation can be prevented through preventing the discharge of polluted water into crop land and 16.60 per cent of them hold the view that industrial purpose sucking ground water could be stopped to prevent land degradation. Of the total 300 respondents 14.00 per cent of them state that land degradation can be prevented through proper disposal of industrial solid waste and 18.67 per cent of them emphasize that land degradation can be prevented through proper treatment of industrial effluents. Moreover, 33.00 per cent of them argue that land degradation can be prevented through proper treatment of industrial wastewater. More than one third of the households of Nagamangalam (40.00%) and Gandhikuppam (36.00%) state that land degradation can be prevented through proper treatment of industrial wastewater.

Data presented in Table 6 indicate the village wise respondents realization of drinking water pollution. It could

Table 4
Village Wise Respondents' Views on Effects of Land Degradation on Living Beings

Village	*Bad odour and vomiting sensation*	*Skin disease*	*Live stocks develop various diseases*	*Disease infection due to accumulation of polluted water*	*Total*
Jagadevi	9 (18)	11 (22)	10 (20)	20 (40)	50
Nagamangalam	7 (14)	9 (18)	10 (20)	24 (48)	50
Malapatti	8 (16)	11 (22)	12 (24)	19 (38)	50
Gandhikuppam	10 (20)	12 (24)	13 (26)	15 (30)	50
Duvarapalli	13 (26)	10 (20)	15 (30)	12 (24)	50
Venkatapuram	5 (10)	13 (26)	17 (34)	15 (30)	50
Total	52 (17.33)	66 (22.00)	77 (25.67)	105 (35.00)	300

Source Computed: Figures in parentheses denote row percentage.

Table 5
Village Wise Respondents' Views on Alternative Ways of Preventing Land Degradation

Village	*Prevention discharge of polluted water on crop land*	*Reducing the level of sucking ground water for industrial purpose*	*Proper disposal of solid waste*	*Proper treatment of industrial effluents*	*Proper treatment of industrial waste water*	*Total*
Jagadevi	10	8	6	9	17	50
	(20)	(16)	(12)	(18)	(34)	
Nagamangalam	9	9	5	7	20	50
	(18)	(18)	(10)	(14)	(40)	
Malapatti	10	11	6	10	13	50
	(20)	(22)	(12)	(20)	(26)	
Gandhikuppam	8	8	8	8	18	50
	(16)	(16)	(16)	(16)	(36)	
Duvarapalli	8	5	11	12	14	50
	(16)	(10)	(22)	(24)	(28)	
Venkatapuram	10	7	6	10	17	50
	(20)	(14)	(12)	(20)	(34)	
Total	55	48	42	56	99	300
	(18.33)	(16.00)	(14.00)	(18.67)	(33.00)	

Source Computed: Figures in parentheses denote row percentage.

be noted that out of the total 300 respondents 28.33 per cent of them realize the drinking water pollution in terms of changes in colour and iron content in the SIPCOT area. More than half of the households of Venkatapuram village (60.00%) observe the water pollution in the form of changes in colour and iron content. This happens because of accumulation of pollutants in groundwater.

In this study, 13.00 per cent of the households observe the drinking water pollution in the form of salty taste and accumulation of impurities and 15.0 per cent of them refer to it as accumulation of sedimentation on some occasions. Moreover, 43.67 per cent of the households realize the drinking water pollution in the form of occurrence of suspended particles in the water. The households of Nagamangalam Village (58.00%), Gandhikuppam Village (58.00%), Malapatti Village (50.00%), and Duvarapalli Village (50.00%) observe the pollution of drinking water in the form of occurrence of suspended particles in the water.

Data presented in Table 7 indicate the village wise respondents' reasons for pollution of drinking water. It could be noted that out of the total 300 respondents 32.00 per cent of them attribute the reason for pollution of drinking water in terms of discharge of chemicals and wastewater into the public water bodies. The households of Duvarapalli village (42.00%) mainly state this reason. It is observed from the study, that 17.00 per cent.

Of the households state that discharge of toxic chemicals by the SIPCOT industries to the adjoining water bodies lead to the pollution of drinking water. It could be noted that 22.67 per cent of the households state that discharge of industrial wastewater and household wastewater into the public water bodies results in the pollution of groundwater. Moreover, 28.33 per cent of the households hold the view that penetration of industrial effluents to the groundwater lead to pollution of drinking water. A considerable number of households of Gandhikuppam village (44%0 and Nagamangalam village (30%) hold the view that penetration of industrial effluents to the groundwater lead to pollution of drinking water.

Table 6
Village Wise Respondents' Realization of Pollution of Drinking Water

Village	*Changes in colour and iron content*	*Salty taste and accumulation of impurities*	*Sometimes sedimentation*	*Suspended particles*	*Total*
Jagadevi	20 (40.00)	6 (12.00)	6 (12.00)	18 (36.00)	50
Nagamangalam	9 (18.00)	5 (10.00)	7 (14.00)	29 (58.00)	50
Malapatti	10 (20.00)	7 (14.00)	8 (16.00)	25 (50.00)	50
Gandhikuppam	8 (16.00)	8 (16.00)	5 (10.00)	29 (58.00)	50
Duvarapalli	8 (16.00)	5 (10.00)	12 (24.00)	25 (50.00)	50
Venkatapuram	30 (60.00)	8 (16.00)	7 (14.00)	5 (10.00)	50
Total	85 (28.33)	39 (13)	45 (15)	131 (43.67)	300

Source Computed: Figures in parentheses denote row percentage.

Table 7
Village Wise Respondents' Reasons for Pollution of Drinking Water

Village	*Discharge of chemical and waste water into water bodies*	*Discharge of toxic chemicals into the public water bodies*	*Discharge of industrial waste and households waste into the public water bodies*	*Penetration of effluents in the ground water*	*Total*
Jagadevi	15	9	14	12	50
	(30)	(18)	(28)	(24)	
Nagamangalam	14	7	14	15	50
	(28)	(14)	(28)	(30)	
Malapatti	12	10	16	12	50
	(24)	(20)	(32)	(24)	
Gandhikuppam	16	7	5	22	50
	(32)	(14)	(10)	(44)	
Duvarapalli	21	8	11	10	50
	(42)	(16)	(22)	(20)	
Venkatapuram	18	10	8	14	50
	(36)	(20)	(16)	(28)	
Total	96	51	68	85	300
	(32.00)	(17.00)	(22.67)	(28.33)	

Source Computed: Figures in parentheses denote row percentage.

Table 8 presents data on the village wise respondents' views on drinking polluted water. It could be noted that out of the total 300 respondents, 21.67 per cent of them realize the effect one drinking polluted water in the form of frequent occurrence of dysentery and diarrhea and 72.33 per cent of them realize the effect on drinking polluted water in the form of frequent occurrence of vomiting and dysentery. It is observed that 12 per cent of the households state that effect of drinking polluted water results in frequent occurrence of occasional occurrence of jaundice and cholera. Moreover, 49 per cent of the households state that the effect of drinking polluted water results in occasional occurrence of cold and related diseases. Half of the households of Jagadevi, Malapatti and Venkatapuram villages realize the effect of drinking polluted water in the form of Occasional occurrence of cold and related diseases.

Data presented in Table 9 indicate the village wise respondents' mode of realizing air pollution in consequence of SIPCOT industrial activities. It could be noted that out of the total 300 respondents, 16.33 per cent of them realize the effects of air pollution in the form of occurrence of bad dour and development of vomiting sensation and 15.33 per cent of them observe it in the form of occurrence of fatigue and headache. It is observed from the data in table that 18.67 per cent of them realize the presence of air pollution in the form of development of eye itching and the rest of them (49.67%) realize the effect of air pollution in the form of difficulty to breathe when they inhale the toxic gases. Half of the Jagadevi village (50%), Gandhikuppam village (54%), and Venkatapuram village (54%) realize the effect of air pollution in the form of difficulty to breathe when they inhale the toxic gases.

Pollution of water bodies around the industrial area leads to various diseases, particularly, skin diseases. It could be noted that increase of toxic elements content of water causes serious damage to biota living within it. Pollution of water causes three environment damages as it harms human health, reduces economic productivity and leads to loss of amenities.

Table 8

Village Wise Respondents' Views on the Effect of Drinking Water Pollution

Village	Frequent occurrence of dysentery and diarrhea	Frequent occurrence of vomiting and dysentery	Occasional occurrence of jaundice and cholera	Occasional occurrence of cold and related diseases	Total
Jagadevi	10 (20)	8 (16)	6 (12)	26 (52)	50
Nagamangalam	12 (24)	9 (18)	5 (10)	24 (48)	50
Malapatti	8 (16)	8 (16)	7 (14)	27 (54)	50
Gandhikuppam	16 (32)	7 (14)	5 (10)	22 (44)	50
Duvarapalli	11 (22)	9 (18)	7 (14)	23 (46)	50
Venkatapuram	8 (16)	11 (22)	6 (12)	25 (50)	50
Total	65 (21.67)	52 (17.33)	36 (12.00)	147 (49.00)	300

Source Computed: Figures in parentheses denote row percentage

In the study area the households and their members are frequently exposed to polluted water bodies. As a result, they are the victims of various types of skin diseases. Out of the total 300 households, 37 per cent of them spend upto Rs. 500 to heal their skin diseases and 32.67 per cent of them incur skin disease treatment expenditure in the range of Rs. 500 - 1000. 30.33 per cent of them spend Rs. 1000 and above per year for skin disease treatment, indicating such households are the major victims of water pollution in the SIPCOT industrial region. The households of Gandhi Kuppam village and Malapatti village relatively spend more amount towards treatment for skin diseases in consequence of water pollution than those of others.

A study of data in Table 10 indicates the Village wise respondents' coping with strategies to protect themselves against environmental degradation. It could be noted that out of the total 300 households, 18.33 per cent of them state that they can cope with environmental degradation in the form of water purification practice along with monitoring public water bodies with respect to level of accumulation of pollutants and 20.33 per cent of them adopt coping mechanism of industrial pollution in the form of monitoring discharge of effluents and other pollutants from each industry. It is significant to note that 20.67 per cent of them practice coping mechanism in the form of conflict with industrialists and making proper representation against industrial pollution to the authorities of pollution control board and government agencies and 16.67 per cent of the households practice coping mechanism in the form of using protective materials. Moreover, 24 per cent of households practice coping mechanism against industrial pollution in the form of undertaking preventive health care measures and practices. Majority of the households of Nagamangalam village (48%) undertake preventive health care measures and practices as coping mechanism against industrial pollution. A considerable number of households of Jagadevi village (32%) and Duvarapalli village (30%) practice coping mechanism in the form of conflict with industrialists and making representations against industrial pollution to the authorities of pollution control board and government agencies.

CONCLUSION

It could be seen clearly from the above discussion that the households rate first order negative externality of starting SIPCOT industrial region in the form of loss of traditional land and household displacement, externality in the form of property damage and development of household behavioural change in consequence of industrial pollution the second, externality in the form of starting SIPCOT industrial region in Bargur in the form of development of conflict between local residents and industrial management the third and development of conflict between local residents and pollution control board the last. It is observed that households rate it as the first order the effect of land degradation in terms of transmission of parasitic diseases through accumulated polluted water, health problem to the live stock in consequence of land degradation the second, occurrence of skin diseases among human beings the third and occurrence of bad odour and vomiting sensation the last.

It is significant to note that households rate it as first order alternative way of preventing land degradation in the form of proper treatment of industrial wastewater, proper treatment of industrial effluents the second, prevention of discharge of polluted water into the crop land the third, reducing the level of sucking ground water for industrial purpose the fourth and proper disposal of industrial solid waste the last.

It is observed from the findings of the study that households practice the first order of coping mechanism against industrial pollution in the form of undertaking preventive health care measures and practices, conflict with industrialists and representation to pollution control board and government agencies the second, coping mechanism of monitoring discharge of industrial waste the third, coping mechanism of water purification and monitoring public water bodies the fourth and coping mechanism of using protective materials the last.

1. The damage to land, river and groundwater is extensive it has been and caused by the industries in the SIPCOT industrial region. The damage to the natural resources has led to an impoverishment of local villagers due to their consequent ill health and the degradation of their livelihood bases. Efforts should be made to ensure that SIPCOT villagers receive full financial compensation for their accumulated losses, including loss of health, ability to work, lost opportunity costs and real livelihood losses due to damage to land and water.
2. The Government must initiate steps to identify the extent and nature of environmental degradation, and begin remediation of the land and water at the expense of the polluting factories in SIPCOT.
3. The villagers of Bargur SIPCOT industrial region have been impoverished by the mounting medical costs, and the inability of people to earn a living due to poor health. The Government should make the industries compensate the villagers for their damaged health, the cumulative medical costs and lost man-days due to poor health.
4. The acute water scarcity, polluted water and air have made life impossible in the SIPCOT region. Villagers, particularly children, suffer from a range of serious health effects. The Government should set up immediate and longterm health monitoring programs, starting with a comprehensive health survey. It should also make the industry pay for long-term medical rehabilitation, through medical facilities that are accessible to and for the villagers.

Rather than creating jobs, industrialisation has robbed many local people of their livelihoods by degrading their livelihood bases, and seriously compromised their ability to work for a living due to health problems. The industry and Government should pay for the initiation of employment training and provide sustainable livelihood opportunities as part of an economic rehabilitation program.

REFERENCES

Taylor Dorceta E. (1992), "The Environmental Justice Movement: No Shortage of Environmental Volunteers," *EPA Journal*, 18(1): 23-25.

Godsil Rachel D. (1991), "Remedying Environmental Racism," *Michigan Law Review*, 90(2): 394-427.

Grossman Karl (1991), "Environmental Racism," *Crisis 98*, 4(14-17): 31-32.

Godsil Rachel D. (1991), "Remedying Environmental Racism," *Michigan Law Review 90*, (2): 394-427, 1.

Mann Eric. (1991), *L.A.'s Lethal Air: New Strategies for Policy, Organizing and Action*. Los Angeles Labor/Community Strategy Center, 80.

Byers R. Lee. (1991), "Regulatory Barriers to Pollution Prevention." *Air & Waste* (Journal of the Air and Waste Management Association), 41(4): 418-422. [I.C]

Carpi John (1994), "Green Batteries: Powering Innovation," *E Magazine*, 5(2): 46–47, 67. [IV]

Casler Stephen D. (1991), "Budget Reallocation and the Peace Dividend: Energy and Pollution Tradeoffs." *Energy Policy*, 19: 666-669. [I.C]

Geiser Ken (1991), "The Greening of Industry," *Technology Review*, 94: 64-72. [III.A]

Holmes Hannah (1992), "Eating Low on the Food Chain," *Garbage*, 4(1): 32-37. [IV]

Lai On-Kwok (1993), "Making Sense of the Greening of Consumption and Production." *Journal of Cleaner Production*, 1(1): 43-47. [III.C]

Mitsch William J. (1993), "Ecological Engineering." *Environmental Science and Technology*, 27(3): 438-445. [II.A]

Piasecki Bruce, and Peter Asmus (1990). *In Search of Environmental Excellence: Moving Beyond Blame*, New York: Simon & Schuster/ Touchstone, [III.A]

Künzli N., Lurmann F., Segal M., Ngo L., Balmes J., and Tager I. B. (1997), Association Between Lifetime Ambient Ozone Exposure and Pulmonary Function in College Freshmen – Results of a Pilot Study, *Environmental Research*, 72: 8-23.

Frischer T., Studnicka M., Gartner C., Tauber E., Horak F., Veiter A., Spengler J., Kühr J., and Urbanek R. (1999), Lung Function Growth and Ambient Ozone: A Three-Year Population Study in School Children, *Am J Respir Crit Care Med*, 160: 390-396.

Schindler C., Künzli N., Bongard J. P., Leuenberger P., Karrer W., Rapp R., Monn C. (2001), Ackermann-Liebrich U., and The Swiss Study on Air Pollution and Lung Diseases in Adult Investigators. Short-Term Variation in Air Pollution and in Average Lung Function Among Never-Smokers: The Swiss Study on Air Pollution and Lung Diseases in Adults (SAPALDIA), *American Journal or Respiratory and Critical Care Medicine*, 163: 356-361.

8

ENVIRONMENTAL POLLUTION: IMPACT ON HEALTH AND ECONOMY-AN OVERVIEW STUDY

Lal B. Suresh , Ravinder. M

INTRODUCTION

The earth is a victim of on slaughter of a materialistic civilization and industrialization. As a result there is rapid depletion of natural resources. Life on the earth being posed to severe threats, nature has started sending us warning signals in the form of droughts, floods, tsunamis, vast changes in climatic patterns, global warming, acid rains, sprawling desertification, depletion of atmospheric ozone shield and many hitherto unheard diseases.1 In the last 100 years mankind managed top destroy much of what took nature millions of years to create on the earth. At no point of time in the history of our planet has so much damage been done. Mankind has poisoned the river systems through out the world and this, in turn, has upset the ecological balance of the rivers and oceans. Trees are indiscriminately being chopped down because of the greed of man.2 Our enormously productive economy demands that we make consumption our way of life, that we convert the buying and use of goods into rituals, that we seek our spiritual or ego satisfaction in consumption. We need things consumed, burnt up, worn out, replaced and discarded at an even increasing rate.3 John Bellamy Foster points out that "the literacy of ecological complaints plaguing the world today encompasses a long list of urgent problems. These include: over population, destruction of ozone layer, global warming, extinction of species, loss of genetic diversity, acid rains, nuclear contamination, tropical deforestation, the elimination of claiman forests, wet land destruction, soil erosion, desertification, floods, famines, the despoliation of lakes streams and rivers, the drawing down and contamination of ground water, the pollution of coastal waters and estuaries, the destruction of coral reefs, oil spills, over fishing, expanding land fills, toxic wastes, the poisonous effects of pesticides and herbicides, exposure to hazards on job, urban congestion and the depletion of non-renewable resources " are to name few.[4]

The above facts tell about the over consumerism of man kind regardless of environmental degradation resultant of rapid exploitation of resources to satisfy greed. The present disastrous environmental degradation is the result of excessive human action in exploiting natural resources to achieve rapid economic development. This grim reality once again necessitates and to refresh our understanding of environment in relation to economic development, severity of environmental degradation and impact of environmental degradation on health and economy.

OBJECTIVES

The major objectives of this paper are as follows:

1. To examine the impact of environmental pollution on health and economy in present day.
2. To focus environmental pollution on health and economy (in term of air and water pollution, land degradation and bio-diversity losses).

DATA AND METHODOLOGY

We have collected the required data and information from various secondary sources like International Geographical Union (IGU) reports, Teri Energy Data Directory Year Book-2007, Economic Survey Reports, 2005-08 (GOI) Ministry of Finance. In order to examine and estimate the impact of environmental pollution on health and economy and various economic as well as health issues.

ENVIRONMENT AND ECONOMY

The study of relationship between environment and economy enables us better comprehend the impact of economic development on environment and economic impact of environmental degradation. Environment is our basic life support system. It is a complex mechanism having two way relationships with man. In one way it indicates nature's influence on man land on the other man's on nature. The natural factors-land form, climate, vegetation, minerals, water, air etc together influence man. Human factors-demography, economy, culture, social attitudes, institutions etc influence the nature. In this way environment is a broad concept comprising diverse surroundings in which we are part perceive, experience and react to events and changes. Economy consists of economic agents involved in economic activities-production and consumption activities.

The relationship between environment and economy can be understood in three ways. Firstly, the environment provides the economy with raw materials to carry out production activities. Secondly it provides life support services such as supply of air, water. Thirdly, it acts as sink for all the waste products that are generated in the process of production and consumption. It is not a passive sink but acts upon the wastes to lean up waste in to useful materials. This capacity of the environment to absorb wastes and clean up is not unlimited. When the wastes generated exceed this capacity of environment, the efficiency of the environment to provide to the economy with goods and services degrades. This is the interference with the nature's delivery of goods and services to human society caused in course for production and consumption activity in the economy.

The level of production of consumption activity is decided through economic interaction. Once the level is decided, the level of interference or obstacles to nature's delivery of goods and services and the damages there of also get decided. If a factory produces some good, it also produces smoke. The amount of goods it produces is decided by the economy. This in turn decides the amount of smoke that will be belched out and the damage to human health it will course.

Thus all the problems of environmental pollution or degradation occurs as a by-product of our activities related to production and consumption that in turn interfere with nature's delivery of raw materials and life support services to the society and it's economy. In view of the relationship between environment and economy, it can be understand that it is the economic activities that cause environmental degradation.

Severity of Environmental Pollution: In the modern era, the development process is more oriented towards industrialization and modernization of agriculture. The environment and development has not been keeping pace with each other. Industrialization and the technology adopted for industrialization have completely misbalanced the environment and ecology. The rapid economic development resulted in the large size increase of pollutants in liquid, solid, gaseous and radioactive forms that cause a multitude of environmental problems. The main environmental problems, various causes of the problems and severity of those problems can be summarized as follows.

Air Pollution

History of air pollution could be traced back to the industrial revolution and the discovery of steam engine in the eighteenth century. Increased use of coal and the smoke and sulfur compounds emanating from it began to contaminate the atmosphere more and more. Air pollution means "presence of pollutants in atmosphere in concentration sufficient to be injurious to human, plant, animal life or property around."[5]

These pollutants include gases, particulate matters, radioactive substance etc. the sources of air pollution are two types–Natural and Man made. Air pollution from human activity on large scale is relatively recent development. Since the industrial revolution, the major fuels have been coal and petroleum. Coal is largely carbon, which, when burns completely produces carbon dioxide, CO2. Petroleum consists largely of hydrocarbons (compounds of hydrogen and carbon), which too, when burn completely form Co2 and water. Sulfur and Nitrogen are essential elements of life and since coal and oil are derived from living organisms, they produce So2, No2 which cause air pollution. Increased use of chemical fertilizers, pesticides is also causing air pollution. Continued use of obsolete chemicals is a dangerous aspect of pesticides produced, used in India, use of obsolete pesticides life DDT, Lindan is also a serious cause of air pollution.

The industries producing pesticides also causing air pollution with inefficient safety measures. The worst example of this kind is Bhopal gas tragedy that killed about over 25,000 people. Coal mining activity is another major source of air pollution. The ambient air quality in the coal washeries is severely degrading due to coal cleaning activities. Suspended Particulate Matters (SPM) in the atmosphere in these areas is 324 mgm per cubic meter as against normal range of 200mgm. The content of No2, Co2 ranges 98% - 101% per cubic meter as against normal range of 80%.[6]

Thus with rapid growth of industrial energy and transport sectors fueled by coal and petroleum and activities involved in producing these fuels are degrading air quality at an increasing rate.

Water Pollution.

Water is an essential and vital component of our life support system. Water pollution can be defined as alteration in physical, chemical or biological characteristics of water making it unfit for designated uses in its natural state. Because of its property to dissolve many substances in it, it can easily be get polluted. Effluents discharged from industries, power plants, coal mines, offshore oil wells, surface run off from agriculture fields, overflowing of small drains, rain water sweeping roads and fields atmospheric deposition etc are some main sources of water pollution. Sewage water, industrial water containing toxic chemicals, acids, alkalis, metallic salts, phenols, cyanides, ammonia radioactive substances, synthetic detergents used in washing and cleaning, Agro chemicals etc are mainly responsible for surface water pollution. In India ground water is used extensively. A variety of land and water based human activities are causing pollution of this precious resource. Septic tanks, deep well injections, mining etc are responsible for ground water pollution. Ground water quality is being increasingly threatened by agriculture, urban and industrial wastes which leads or ejected into underlying aquifers. It has been established that once pollution has entered the surface environment, it may remain concealed for years being dispersed over wide areas of ground water aquifers and reducing ground water quality.

A wide variety of both inorganic and organic pollutants are present in effluents from breweries, tanneries, dying textiles, paper and pulp mills, steel industries, mining operations etc. These cause serious ground water pollution. Modern agriculture is heavily relying on wide range of chemical fertilizers and pesticides. Nitrates fertilizers used on soil enter wells causing this water nitrates rich. The pollution of surface water and ground water is very much severe in and around large cities and industrial areas. A variety of industries located on the banks or catchment areas of main rivers in India. Pollution River waters to such an extent that any attempt to control the pollution will have no significant effect. The incidence of fluoride above permissible levels of 1.5 ppm occurs in 14 Indian states- Andhra Pradesh, Bihar, Gujarat, Haryana, Karnataka, Kerala, Madhya Pradesh, Maharashtra, Orissa, Punjab, Rajasthan, Tamilnadu, Uttar Pradesh and West Bengal. High levels of salinity are reported from all these states except West Bengal. High levels of arsenic above permissible levels of 50 parts per billion (ppb) are founds in the alluvial plains of the Ganges covering six districts of West Bengal. Presence of heavy metals in ground water is found in 40 districts from 13 states. Nitrate concentration is above the permissible level of 45 ppm in 11 states, covering 95 districts. A survey under taken by Central Pollution Control Board (CPCB) in 1995 identified 22 sites in 16 states as critical for ground water pollution caused by industrial effluents. Traces of heavy metals such as lead, cadmium, zinc and mercury were reported by the survey under taken by Centre for Science and Environment (CSE) in eight places in Gujarat, Andhra Pradesh and Haryana. A shallow aquifer in Ludhiana city, the only source of its drinking water, is polluted by a stream which receives effluents from 1300 industries. Excessive withdrawal of ground water from coasted aquifers has led to induced pollution in the form of sea water intrusion in Kachch and Saurashtra in Gujaraat, Chennai in Tamilnadu and Kozhikode in Kerala.[7]

Land Degradation

Soil erosion is the main cause of land degradation. Soil is the upper layer of the earth which is formed by weathering of rocks. In India, nearly 200 million hectares of land out of 329 millions hectares was eroded due to floods, cyclones and deforestation. During 1901-1972 forests in the area about 3,402 thousands hectares were destroyed for the building of river valley projects, transport, communication and other industrial units. Effluents discharged by various industrial units in liquid, solid and gaseous forms, fly ash from thermal power plats and ash ponds erected by them, mining and granite activities, drainage water, municipal wastes, computer wastes, wastes generated by hospitals etc indiscriminate use of chemical fertilizers and pesticides in modern agriculture all these are responsible for land degradation. Construction of heavy water storage dams and over irrigation of fields are resulting in salinization of soils where new cultivation methods (Green Revolution) are being extensively used.

Bio-Diversity Degradation

Bio-diversity degradation is one of the threatening issues of environmental problems. The "earth summit" At Rio De

Janeiro in 1992 sensitized the world about this problem and recognized the need to name huge number of species and mapping bio-diversity in view of fast extinction many of the species. The tropical rain forests are main habitats of millions of species of plants, birds, amphibians, insects as well as mammals. About 50-80% global bio diversity lies in these rain forests. India has a rich biological diversity of flora and fauna, considered to be one of the 12 mega-bio-diversity countries in the world. Though extinction or elimination of a species is a natural process of evolution, the process of extinction has become fast in the current years of human civilization if the present trend continues 1/3rd to 2/3rd of current bio-diversity would be lost by the middle of the 21st century. Destruction and loss of natural habitat is the single largest cause of bio-diversity loss. All the economic development activities directly or indirectly are responsible for bio-diversity degradation.

POLLUTION IMPACT ON HEALTH AND ECONOMY

In the present age of technological revolution, the needs of human beings are increasing rapidly. As a result, irrational and intensive exploitation of national resources is being taken place leading to global warming, ozone depletion, increasing amounts of CFCs, acid rains. Due to this, misbalancing in the Eco system is taking in the form of disastrous floods, earth quakes, landslides epidemics etc. Environmental degradation can lead to a self-perpetuating process due to economic necessity and economies may destroy or exhaust the resources to continue economic activities for survival. Thus pressure on the natural resources ends up in environmental degradation. Environmental degradations have severe consequences for self-sufficiency, income distribution and future development potential. The economic impact of Environmental Pollution can be known by analyzing spread of diseases and consequent rise in health expenditures, reduced productivity of resources, falling farm productivity loss of bio diversity, live stock degradation corrosion of materials, increasing industrial costs due to Environmental

Pollution. Economic development in any region is supposed to increase employment opportunities, incomes and living standards of the people of that region. Development also leads to pollution in areas where it takes place and when pollution control measures are not properly implemented. Pollution levels reach to disastrous state where the development does not create employment and income opportunities, instead leads to decline in the employment, income and living standard of the people. Industrial activities cause severe air, water, soil pollutions, ultimately cause to decline in employment, income and living standards of people, which are prime determinants of the health status of the people and ends up in the degradation health status of people.

Air is essential ingredient for all living beings on the earth. Man has polluted the atmosphere so heavily that the much of the population now breaths a mixture of highly toxic gases with every lungful of air. Human body requires approximately 25 kg of air per day to sustain its requirement of oxygen. As far as food consumption is concerned, it is estimated that 1.5 kg of food is required per day per person. That means that man inhale 15-20 times higher quality of air than the quality of food intake. This explains that pollutants concentration in Air should be lesser than the pollutant concentration in food.

Various pollutants which pollute air have harmful effects on human health and other living beings in the biosphere; carbon monoxide (CO) is very harmful, causes headache and irritation mucous membrane. It combines with hemoglobin of blood and reduces its O2 carrying capacity. At about 200 ppm for 6-8 hours, there begins head ache, above 300 ppm begins throbbing head ache followed by vomiting and collapse, at above 500 ppm man reaches in to coma and at 1000ppm, there is death. SO2 causes intense irritation to eyes and respiratory tract. Plants are relatively more sensitive to SO2, than are animals and Man. In most plants leaf area collapses under intense exposure to SO2. There is bleaching of leaf pigments due to conversion of CHL-a to phaeophytin-a. Thus So2 exposure has any impact on plant productivity. It is toxic to plants. Acid rains caused by SO2 create complex problems, they increase soil acidity, thus affect land flora and fauna, cause acidification of lakes and streams there by affect aquatic life, affects crop productivity and human health, corrode buildings, bridges. Due to acidity levels of heavy metals increase beyond safe limits. Thousands of lakes in USA, Canada, and Norway have become unproductive due to acidity. Fish population decreased tremendously.

The acid rain is fast spreading to developing world where tropical soils are more vulnerable than those n Europe. Due to SO2 from coal based thermal power plants and petroleum refinery pH value of rain water is below the critical level in Delhi, Nagapur, Pune, Bombay and Calcutta. Increase in O3 (Ozone) concentration near the earth's surface reduce crop yields significantly, has adverse impact on human health. In California, U.S.A fruits and vegetables yields have reduced due to O3 pollution. In U.S.A. air pollution causes a crop loss worth 2 billion dollars. Ozone is alone and in combination with SO2 and NO2 causing GNP losses of 50% in several European countries. In Denmark O3 is affecting potato, clover, spinach, alfalfa etc. On the other hand depletion of ozone layer in stratosphere increase UV radiation on earth results in many skin diseases including cancer to human beings. Destruction of aquatic life and vegetation and loss of immunity are disastrous of effects of ozone depletion.8 Increase in the green house gases in the atmosphere, particularly CO2, gradually warming the earths atmosphere is termed as global warming or green house effect that is harmful to the environment and human health and biodiversity. The green house gases- CO2, Methane, CFCs, NO and O3, which are main cause of global warming, present in the atmosphere in small amount but play major role in heat balance of atmosphere and bring changes in climate. Evidences show that there are abrupt changes in the climate due to rapid increase in green houses gases. The Inter Governmental Panel on Climate Change (IPCC) has concluded that forests are highly sensitive to climate change and that up 1/3 of current forested areas could be affected in some way. As forests harbor most of the biodiversity, this indicates increasing threat to biodiversity.

Due to climate change summer mortality in US has been estimated to rise from current total around of 1150 deaths to around 7450. A number of non-parasitic communicable diseases including yellow fever, dengue and other arbovirus diseases, plague, dysentery and other diarrhoeal conditions may spread. The patterns of development and multiplication of malaria parasites or viruses with in these mosquito hosts is temperature dependant. If temperature increases, Egyptian water snails may be able to spread schistosomiasis all the year round; increasing the already heavy parasitic burden of the Egyptian peasantry. The major mosquito-borne diseases in Australia are dengue fever are result of green house effect.9 Thus with the climate change there will be an increasing number of disasters which adversely affect health and Economy with increasing vulnerability of people to diseases and decreasing productivity in agriculture and industries. Sea levels will rise by 18 cm by 2040 due to global warming. Although the world overall will become more humid, some areas will become drier. Most areas of the world that currently suffer from drought are predicted to become ever drier expected doubling in the annual number of hot days in temperature zones will affect people who are vulnerable to extreme heat such as very young, very old and those having chronic respiratory diseases, heart diseases or high blood pressure. Global warming extends range of insects carrying diseases such as malaria in to regions where people have developed no immunity. Animal health could also be affected by the spread of diseases by parasites. Food production in temperate areas will probably affected by the attract of insects that in the past have been killed off in large measures during the winters but that could survive and flourish under warmer conditions.10 Thus with climate change there will be an increasing disasters which adversely affect health and economy, with increasing vulnerability of people to diseases and health expenditure and decreasing productivity in agriculture and industries.

Water pollution adversely affects the health of people. Animals and plants and pollutes the soil and thereby reduces productivity of labor, animals and plants causing decreasing yields in agriculture. Water pollution also adversely affects industries by increased costs of inputs polluted water bodies are harmful and are vehicles of many harmful diseases as cholera, dysentery, typhoid etc. The polluted water supplies are responsible for more human illness than any other environmental influences. Nearly 80 percent of world's diseases more so in developing world, can be linked with water.

Sewage and domestic wastes are excellent medium for the growth of pathogenic bacteria, viruses and protozoa. Domestic sewage that mainly composed of spent water containing wine, soapy water, food materials, make water completely unfit water for drinking and domestic use. Micro organisms in polluted water cause chronic diseases in man and animals. Water polluted with industrial effluents causes kidney, liver, brain and lungs diseases. Accumulation of nitrates, due to excess application of nitrogen fertilizers to the soil, in water cause, when cattle and man drink this water, damage to respiratory and vascular system, because of which suffocation and death may cause. Thermal pollution in water decreases dissolved oxygen and increases biological Oxygen demand (BOD). Decrease in solubility of gases in water, causes excessive eutrophication, rapid setting of sediment load affecting aquatic food supply.[11]

In Marine water, the serious pollutant is oil that causes damage to marine fauna and flora including algae, Fish, birds. Hydrocarbons and benzpyrene accumulate in food chain and consumption fish by man may cause cancer, Detergents used to cleanup the oil spills are also causing harm to marine life. Mercury enters water in organic and inorganic forms through industrial effluents are highly poisonous. Mercury was responsible for the Mina Mata epidemic that caused several deaths in Japan and Sweden. Lead processing industries discharge lead in to water. Lead pollution causes damage to liver and kidney reduction in hemoglobin formation, mental retardation and abnormalities in fertility and pregnancy. Fluoride present in water and soil besides air. Cattle grazing around fluoride sources develop fluorosis. The toxic effects are staining, mottling and abrasion of teeth, high fluoride levels in bone and urine, decreased milk production and lameness. Symptoms in human beings are yellowing teeth, damaged joints and bone deformities. Prolonged exposure and use of contaminated water above the permissible limit of arsenic at 0.05mg per liter set by Bureau of Indian Standard is associated with a gamut of serious illnesses. In many villages, it is common to find entire families suffering from Keratosis, a skin disease on hands and feet in West Bengal. [12]

Continuous use of contaminated water for irrigation of agriculture fields led to soil degradation around major industrial units especially in and around thermal power plants. Soil pollution coupled with air pollution led to decrease in yield and change in the cropping pattern. Due to continuous use of contaminated water, cultivable lands become barrier. Due to water, soil and air pollutions, grazing contaminated grass and drinking polluted water, cattle health was adversely affected. Live stock is more sensitive to pollutants in contrast to man because of toxins of the pollutants directly affects them. Most animals have short life span and rapid physiological activities and respond more quickly and readily to pollutants. There is decline in lactation period and milk yielding in milch animals, cattle rearing was given up due to severity of pollution levels. Degradation in biodiversity, caused by air, water and soil pollution, retards development of all sectors of the economy in multiple ways. Thus, the environmental pollution may adversely affect the health and economy reducing incomes and physical qualities of life of people.

CONCLUSION AND SUGGESTIONS

To day, environmental conservation, ecological balance and prevention environmental pollution concern every citizen. Environment belongs to each one of us and all of us have responsibility to protects from being degraded by excessive human action to satisfy greed. Adoption of western development model which leads to over exploitation of resources and environmental degradation should be rationalized, we need to recognize and identify indigenous economic living systems that are ecologically sound. Many indigenous technologies, skills and processes are found appropriate for sustainable development and harmonious with nature and community. We should not only adopt appropriate technology but also see that appropriate products are produced by this appropriate technology so that the consumption of variety of durable goods, industrial products which lay undue pressure on natural resources is rationalized. Irrational use of natural resources in pursuing development without bothering about the sustainability, possibly affects the interests of future generations in negative way. This is violating the natural rights of future generation to inherit efficient, quality environment which the concept of 'Sustainable Development' emphasizes to be protected. Gandhiji wanted the society not to exploit in excess and advocated optimal utilization of natural resources. So, that the share of future generation in the national resources is not stolen by present generation. The interests of poor, who are natural stake holder of the natural resources also should be safe guarded. Their share in development fruits also should not be stolen due to exclusive development model. The relevance of Gandhian model of economic development should be recognized to over rid the menace of environmental degradation.

REFERENCES

1. Hema Lata Sharma (2006), 'Environmental Conservation–Key to Sustainable Development', University News, March, 20-26, 57p.
2. Jagbir Singh (Editor) (2005), 'Environment and Development, Challenges and Opportunities', I.K.International Pvt. Ltd., New Delhi.
3. Allan Thein (1992), 'During, How Much is Enough?' New York, 21-22pp.
4. John Bellamy Foster (1995), "The Vulnerable Planet: A Short Economic History of Environment, Monthly Review Press, New York.

5. M.R.Garg, V.K.Bansal & N.S.Tiwana et.al, (1995), 'Environmental Protection and Pollution,' Deep and Deep Publication, 26p.

6. The Ennadu (2007), Telugu Daily, December, 25.

7. The Hindu Survey of Environment (2004), 7-8pp.

8. P.D.Sharma (1995), 'Economy and Environment', Rastogi Publication, Meerut, 308-315pp.

9. S.S.Purohit and R.Ranjan (2003), 'Ecology, Environment and Pollution, Agrobibs (India), 739-747pp.

10. S.H.Raza, (S.K.Agarwal, Editor) (2003), 'Global Warming–A Global Warming Environmental Scenario for 21st Centrury', APH Publishing Corporation, New Delhi, 178-179pp.

11. R.K.Khitoliya (2004), 'Environmental Pollution, Management and Control for Sustainable Development', S.Chand, 44-45pp.

12. The Hindu, Survey of Environment (2004), 33p.

9

MODIFICATION OF PARTICLE DEPOSITION MODEL IN TURBULENT PIPE FLOW

Esam Ismaeel Jassim

INTRODUCTION

Particle migration and deposition have taken a considerable attention due to its significance in numerous engineering applications ranging from aerosol air pollution to the microelectronics industry. Other applications including heat exchangers, health effects and chemical industries are of interest in such phenomena as it can cause environmental and health risks.

Because aerosol particle deposition is an important factor in measuring the risk and pollution level, many experimental, empirical and theoretical studies have attempted to understand the process. During past few decades, intensive researches have been conducted to describe the phenomena of the particle deposition in turbulent flow. Schwendiman and Postma (1961), Wells and Chamberlain (1969), Liu and Agarwal (1974), Friedlander and Johnston (1957), Sehmel (1968), and Ilori (1971) reported extensive experimental data for particle deposition rate in turbulent duct flows. In general, the studies concluded that deposition velocity has a V-shaped variation with a minimum at the particle relaxation time of about 0.1 to 0.5 wall units (Chen et al., 1997). The rate of particle deposition increases as the relaxation time decreases in the diffusion regime whereas it increases when the relaxation time increases in the impaction regime.

Theories of particle deposition process have also been addressed by many researchers; Fuchs (1964), Wood (1981), Hidy (1984), Papavergos and Hedley (1984), and Hinds (1999). Semi-empirical correlations for evaluating particle deposition rates in turbulent ducts have been proposed by Friedlander and Johnston (1957), Davies (1966), and Cleaver and Yates (1975); however, the expressions are valid for smooth walls. Further progress in developing the deposition model was reported by Fichman et al. (1988), Fan and Ahmadi (1993, 1994), Chen and Ahmadi (1997), Shams et al. (2000), and Tian and Ahmadi (2007).

Numerical studies of transport and deposition of particles in turbulence flow field, both vertical and horizontal, have been conducted as well in the past few decades. Among lots of literatures studied such phenomena, Li and Ahmadi (1991, 1993) performed a series of numerical simulations on deposition of small particles in a turbulent channel flow. Li et al. (1994) presented digital simulation results for particle deposition rate in an obstructed turbulent duct flow. Finally, Tain and Ahmadi (2007) compared different computational models for predicting particle deposition in the turbulent duct flows.

As a conclusion, earlier and recent works are confirmed that the deposition velocity has a V-shaped variation; however further studies are still needed to tune the correlation in such a shape that can be more consistence with and to precisely address the available experimental data. In this work, a step towards such tuning is presented through performing modification to the model of Fan and Ahmadi (1993) in order to enhance the prediction of the deposition rate.

The study starts with developing and analyzing the mechanism of the motion of particles in turbulent regime ending up with a correlation describes the particle velocity profile for submicron and supermicron particles. Then the definition of the concept of deposition velocity and the revision of the contribution of the literature in such manner are discussed. Followed by proposing the significance of introducing new model to better read and describe the merit of the particle deposition phenomena. Comparison with existing experimental data and other recent empirical expressions is also reported.

The particle diameter ranging from 1nm to 200 ìm are studied. The effects of particle to flow density ratio, pipe size, and roughness are addressed as well. The deposition efficiency, a tool used in this work to measure the deposition rate of particles, is evaluated and compared with that predicted by other models.

PARTICLE MOTION IN TURBULENT FLOW

In an accelerating flow field, particles are not able to follow perfectly the direction of fluid motion due to their own mass. For instance, in straight-line accelerating (decelerating) motion, particles lag behind (fall ahead) the Lagrangian "fluid particles". For submicron particles, and in the absence of external force fields, the Brownian or turbulent diffusion is the basic mechanisms that drive deposition (Crowe, 2006). For particles larger than about 1 ìm, deposition is primarily due to inertial impaction and gravitational settling.

In this section, the mechanism of the motion of particles suspended in a moving fluid is briefly discussed. The effects associated with the particles' own inertia becomes manifested particularly in accelerating fluid motion. The mechanisms driving particle motion are synonymously called deposition mechanisms as the result of such phenomena is to make particles migrate to the bounded surfaces and deposit there.

Since the diameter of the particle is much smaller than the carrier pipe radius and its density is much greater than that of the gas flow, the particle equation of motion in the fully turbulent region can be reduced to an equation consisting only of the drag and gravitational forces (Crowe, 2006).

The general expression for the drag force F_D on a spherical particle in a gas of constant velocity can be written as:

$$F_D = \frac{1}{2}\pi r_p^2 \rho_g \left|u_p - u_g\right|(u_p - u_g)\frac{C_D}{C_c} \quad (1)$$

Where, C_c is the Cunningham non-continuum correction factor, defined as:

$$C_c = 1 + \frac{\lambda}{d_p}\left[2.34 + 1.05\exp\left(-0.39\frac{d_p}{\lambda}\right)\right] \quad (2)$$

The drag coefficient can be estimated from the following formulas depending on the particle Reynolds number for:

$$C_D = \frac{24}{\mathrm{Re}_p} \qquad \mathrm{Re}_p < 0.1 \quad (3a)$$

$$C_D = \left(\frac{24}{\mathrm{Re}_p}\right)\left[1 + \frac{3\,\mathrm{Re}_p}{16} + \frac{9\,\mathrm{Re}_p^2}{160}\ln\left(2\,\mathrm{Re}_p\right)\right] \quad 0.1 < \mathrm{Re}_p < 2 \quad (3b)$$

$$C_D = \left(\frac{24}{\mathrm{Re}_p}\right)\left[1 + 0.15\,\mathrm{Re}_p^{0.687}\right] \qquad 2 < \mathrm{Re}_p < 500 \quad (3c)$$

$$C_D = 0.44 \qquad 500 < \mathrm{Re}_p < 2.10^5 \quad (3d)$$

Analysis of Particle Flow in the Fully Turbulent Region

In most cases, the particles with diameter $d_p < 1$ mm are considered to have low Reynolds numbers (*Stokes flow*) while the particles with d_p ³ 1 mm are considered to have significant Reynolds numbers as suggested by Crowe (2006). Consequently, the modeling approach adopted in the present work can be looked at for the following flow conditions:

Stokes Flow ($d_p < 1$ mm)

Equation of motion becomes:

$$m_p \frac{du_p}{dt} = \frac{3\pi\mu_g d_p}{C_c}\left(u_g - u_p\right) - m_p g \tag{4}$$

Introducing the *particle relaxation time (velocity response time)*:

$$\tau_V = \frac{\rho_p d_p^2 C_c}{18\mu_g} \tag{5}$$

$$\frac{du_p}{dt} = \frac{\left(u_g - u_p\right)}{\tau_V} - g \tag{6}$$

The solution to this equation for constant u_g and a general initial particle velocity u_{p0} is:

$$u_p = \left(u_g - g\tau_V\right)\left[1 - \exp(-t/\tau_V)\right] + u_{p0}\exp(-t/\tau_V) \tag{7}$$

For zero initial particle velocity, Eq. (7) can be simplified to:

$$u_p = \left(u_g - g\tau_V\right)\left[1 - \exp(-t/\tau_V)\right] \tag{8}$$

The terminal velocity of a spherical particle due to gravitational settling is obtained from the following expression (Hinds, 1999):

$$u_s = \frac{\rho_p d_p^2 g C_c}{18\mu_g} \tag{9}$$

or

$$u_s = g\tau_V \tag{10}$$

Thus:

$$u_p = \left(u_g - u_s\right)\left[1 - \exp(-t/\tau_V)\right] \tag{11}$$

Flow with Significant Reynolds numbers (d_p > 1 mm)

Equation of motion becomes:

$$m_p \frac{du_p}{dt} = \frac{3\pi\mu_g d_p C_D \mathrm{Re}_p}{24C_c}\left(u_g - u_p\right) - m_p g \tag{12}$$

Introducing the *velocity response time* and the coefficient $\varsigma = \frac{C_D \mathrm{Re}_p}{24}$ leads to:

$$\frac{du_p}{dt} = \frac{\varsigma}{\tau_V}\left(u_g - u_p\right) - g \tag{13}$$

Solution of equation (13) for constant u_g and a general initial particle velocity u_{p0} becomes:

$$u_p = \left(u_g - g\frac{\tau_V}{\varsigma}\right)\left[1 - \exp\left(-\frac{t\varsigma}{\tau_V}\right)\right] + u_{p0}\exp\left(-\frac{t\varsigma}{\tau_V}\right) \tag{14}$$

For zero initial particle velocity, Eq. (14) becomes:

$$u_p = \left(u_g - g\frac{\tau_V}{\varsigma}\right)\left[1 - \exp\left(-\frac{t\varsigma}{\tau_V}\right)\right] \tag{15a}$$

or:

$$u_p = \left(u_g - \frac{u_s}{\varsigma}\right)\left[1 - \exp\left(-\frac{t\varsigma}{\tau_V}\right)\right] \tag{15b}$$

The Particle Velocity Profile

Particle velocity profiles are evaluated and plotted for different particle sizes. Figure (1) shows the dimensionless mean particle velocity profiles for submicron sizes (1 nm-1mm) as a function of dimensionless distance from the wall at certain time using Equation (8). As shown in the figure, the smaller particles have almost the same velocities as the carrier fluid velocity. The deviation occurs as a result of gravity when the particles grow. In fact, as the particle size increases the particle response to the fluid velocity decreases giving rise to an increase in the relative velocity. In the other words, the time required for the particles to reach (~63%) of the carrier fluid velocity, the relaxation time, increases as the particle becomes larger.

This conclusion can clearly be derived from Figure (2), which illustrates the development of the particle velocity

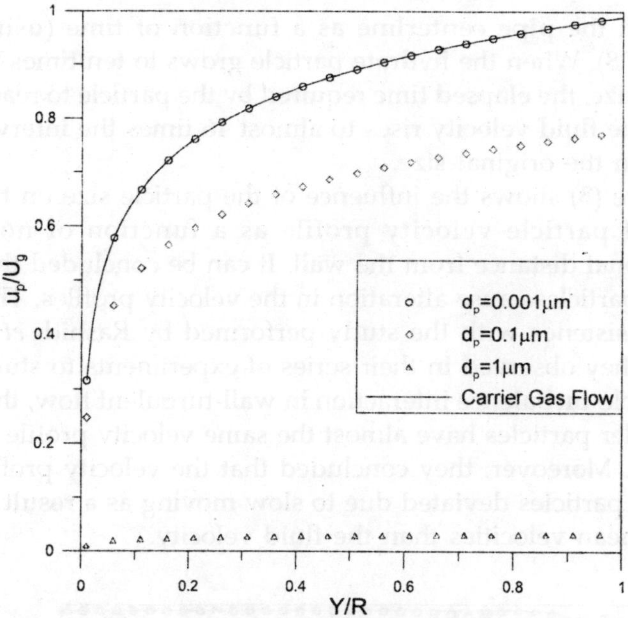

Figure 1: The Velocity Profile of Different Particle Sizes after 0.1μs

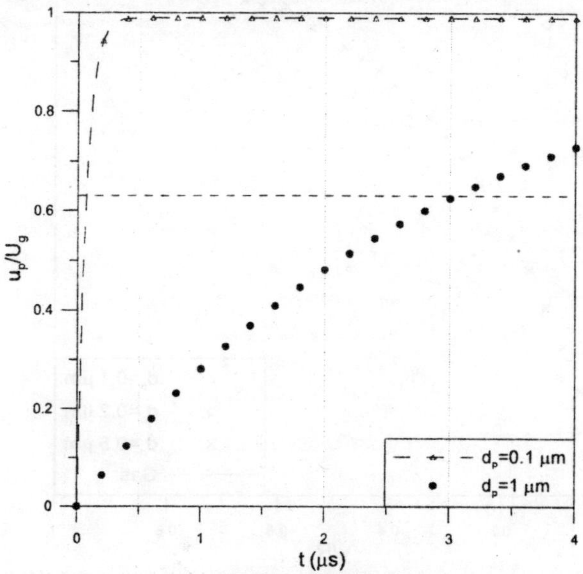

Figure 2: Dimensionless Mean Particle Velocity Profile as a Function of Time

profile at the pipe centerline as a function of time (using Equation 8). When the hydrate particle grows to ten times its original size, the elapsed time required by the particle to reach 63% of the fluid velocity rises to almost 46 times the interval needed in the original size.

Figure (3) shows the influence of the particle size on the shape of particle velocity profile as a function of non-dimensional distance from the wall. It can be concluded that heavier particles cause alteration in the velocity profiles. This is in consistence with the study performed by Rashidi *et al* (1990). They observed in their series of experiments to study the particle-turbulence interaction in wall-turbulent flow, that the smaller particles have almost the same velocity profile as the fluid. Moreover, they concluded that the velocity profile of larger particles deviated due to slow moving as a result of slower mean velocities than the fluid velocity.

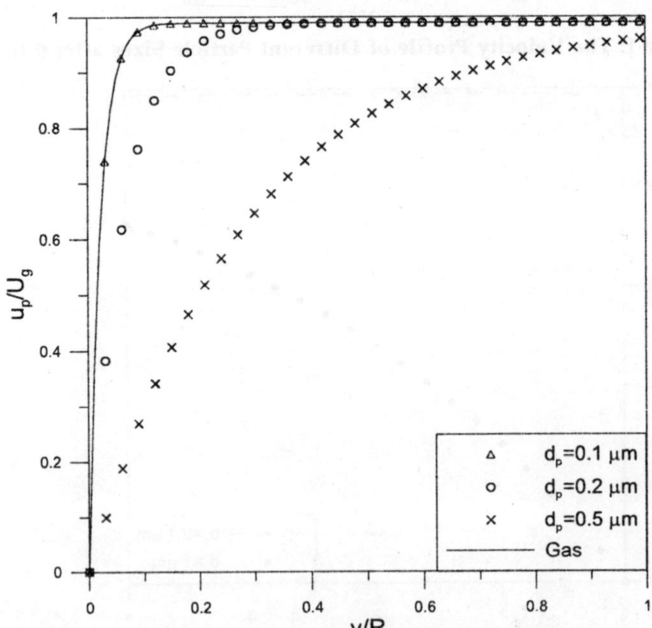

Figure 3: Dimensionless Mean Particle Velocity Profile as a Function of Non-dimensional Distance from the Wall

THE DEPOSITION VELOCITY

Analogous to the settling velocity for deposition by settling, the deposition velocity is the effective velocity with which particles migrate to a surface. The non-dimensional deposition velocity for particles released with undisturbed concentration is given by (Tain and Ahmadi, 2007):

$$V_d^+ = \frac{G_p}{C_{av} u^*} \tag{16}$$

where, G_p is the particle mass flux to the wall and C_{av} the particle average concentration above the surface. The latter is usually taken equal to the average concentration over the cross-section of the conduit (Crowe, 2006).

Numerous correlations for deposition velocity were presented based on experimental data. A simple expression for nondimensional deposition velocity proposed by Wells and Chamberlain (1967) is given by the following:

$$\frac{V_d}{u^*} = 0.2Sc^{-2/3}\,\mathrm{Re}^{-1/8} \tag{17}$$

where Sc is the Schmidt number, defined as:

$$Sc = \frac{\nu_g}{\Gamma} \tag{17a}$$

However this correlation is valid only within the particle size range of submicron. For other regimes, the correlation proposed by Friedlander and Johnstone (1957) is frequently used for calculating the deposition velocity of particles:

$$\frac{V_d}{U_g} = \begin{cases} \dfrac{f/2}{1+\sqrt{f/2}\left(1525/(S_L^+)^2 - 50.6\right)} & S_L^+ < 5 \\ \dfrac{f/2}{1+\sqrt{f/2}\left\{5\ln\left[5.04/(S_L^+/5 - 0.959)\right] - 13.73\right\}} & 5 \le S_L^+ \le 30 \\ \dfrac{f}{2} & 30 < S_L^+ \end{cases} \tag{18}$$

where, f is the friction factor; and

$$S_L^+ = \frac{\rho_g d_p^2 \rho_p U_g^2 f/2}{18\mu_g^2} \tag{19}$$

Other empirical models are that suggested by Wood (1981) and Fan and Ahmadi (1993). The latter is modified in this study to account for flow configuration. The Wood nondimensional deposition velocity is given by:

$$V_d^+ = 0.057Sc^{-2/3} + 4.5\times10^{-4}\tau^{+^2} + \tau^+ g^+ \tag{20}$$

The nondimensional particle relaxation time and the gravitational sedimentation in horizontal flow are respectively given by:

$$\tau^+ = \frac{\tau u^{*2}}{\nu} = \frac{\rho_p d_p^2 u^{*2} Cc}{18\rho_f \nu_g^2} \tag{21a}$$

$$g^+ = \frac{\nu_g}{u^{*3}} g \tag{21b}$$

The first term in Eq. (20) is resulted from the Brownian motion and eddy diffusion, the second term is the consequences of particle deposition by the eddy diffusion-impaction mechanism, while the last term stands for the gravitational sedimentation on the lower wall of horizontal pipes (Chen and Ahmadi, 1997).

Fan and Ahmadi proposed an empirical correlation for deposition of particles including the effects of surface roughness and gravity. For horizontal ducts, the expression takes the form:

$$V_d^+ = \begin{cases} 0.084Sc^{-2/3} + \frac{1}{2}\left[\frac{\left(0.64k^+ + 0.5d_p^+\right)^2}{3.42}\right]^{1/(1+\tau^{+^2} L_1^+)} \\ \times\left[1 + 8e^{-(\tau-10)^2/32}\right]\frac{0.037}{1-\tau^{+^2} L_1^+} + \tau^+ g^+ & \text{if } V_d^+ < 0.14 \\ 0.14 \end{cases} \tag{22}$$

$$\text{where } L_1^+ = \frac{3.08}{Sd_p^+};\ S = \frac{\rho_p}{\rho_f};\ d_p^+ = \frac{d_p u^*}{\nu_g};\ k^+ = \frac{ku^*}{\nu_g} \tag{23}$$

k^+ is the dimensionless surface roughness (which is zero for smooth surface)

Eq.(22) consists of two terms the first term represents the effect of the particle diffusion and is dominated when the particle is in the range of submicron while the second term stands for inertia effect.

Improving of Eq. (22)

Literatures have demonstrated that the local particle deposition relays both on particle size and the actual aerodynamic conditions (Browne, 1973; Miguel et al, 2004; Sosnowski et al., 2007). Thermophoresis via temperature gradients or electrostatic drifting due to charged particles or electric fields near charged surfaces can also influence deposition (Hussein et al., 2009). This paper is focused on the influence of aerodynamic conditions particularly flow Reynolds number.

Experimental results of Miguel et al (2004) showed that deposition efficiency has significant sensitive to the flow Reynolds number. Browne (1973) studied analytically the influence of Reynolds number on the deposition velocity of particles. The results of his analysis showed that significant variation occurs to the particle deposition rate when the flow Reynolds number changes particularly for tiny particles. He also concluded that the variation becomes less as the particle size increases.

Identical conclusion has been obtained in most recent research accomplished by Sosnowski et al. (2007). In particular, they found that the deposition of submicron particles, caused primarily by Brownian diffusion, is strongly influenced by local aerodynamic effects. In turn, large particles are less sensitive to such variation due to their high inertia. Consequently, it is essential to investigate the significance of flow Reynolds number that was dropped from the Fan and Ahmadis' empirical expression of particle deposition.

A modification has applied to the first term of Eq. (22) by involving the effect of flow configuration, as suggested by Browne (1973) and Sosnowski et al. (2007). Wells and Chamberlain correlation, Eq. (7), is a proper formula that describes the behaviour of particle deposition in the Brownian region since it takes into account the convective term of diffusion which emerges in the Reynolds number. Hence, the new expression combining Fan and Ahmadi and Wells and Chamberlain correlations will take the form:

$$V_d^+ = \begin{cases} 0.2Sc^{-2/3}\,\mathrm{Re}^{-1/8} + \dfrac{1}{2}\left[\dfrac{\left(0.64k^+ + 0.5d_p^+\right)^2}{3.42}\right]^{1/(1+\tau^{+2}L_1^+)} \\ \times\left[1+8e^{-(\tau-10)^2/32}\right]\dfrac{0.037}{1-\tau^{+2}L_1^+} + \tau^+ g^+ \text{ if } V_d^+ < 0.14 \\ 0.14 \end{cases} \tag{24}$$

Eq.(24) is the expression to be studied for assessing its validity in predicting the rate of particle deposition.

THE DEPOSITION RATE

A basic quantity characterizing convective transport of a particle is the *penetration ratio*, or *penetration fraction*, defined as the fraction of the inflowing particles that exit the flow system. It can be written as:

$$f_p = \frac{C_{out}}{C_0} \tag{25}$$

Where, C_{out} and C_0 are the particle concentrations (either mass or number) at the outlet and inlet, respectively.

For fully developed turbulent flow in a circular pipe of length L, the deposition velocity V_d may be considered as constant along the tube, and the penetration fraction can be express as (Crowe, 2006):

$$f_p = \exp\left(-\frac{4V_d L}{u_g D}\right) \tag{26}$$

Where, u_g is the carrier gas velocity; D is the pipe diameter.

Alternatively, *Collection efficiency* can be defined as the fraction of the inflowing particles that is lost by deposition:

$$\eta_c = 1 - f_p = 1 - \exp\left(-\frac{4V_d L}{u_g D}\right) \tag{27}$$

The above expression enables the calculation of particle deposition rate through knowledge of the deposition velocity discussed earlier.

COMPARISON OF THE PROPOSED MODEL WITH EXPERIMENTAL DATA AND OTHER MODELS

To appreciate the process of particle deposition predicted by the current model, we compared the theoretical predictions of the model versus some existing experimental data and semi-empirical correlations. Although experimental data for particle deposition from turbulent flows is relatively scarce, there is however more data for deposition from aerosols applications.

The simulation results for particle deposition for turbulent air flow in a 40 mm diameter pipe with 1m length and Reynolds number of 5000 based on the pipe diameter are presented in this section. Air at 270 K ($\mu = 2.0 \times 10^{-5}$ kg m s^{-1}, $\rho = 1.23$ kg m^{-3}) is used in the analysis with the particle to air density ratio of about 750. The mean free path (λ) appeared in Eq. (2) is 62 nm as was calculated by Maxwell (1868).

Figure (4) compares the suggested model with those of earlier studies for horizontal pipe. The experimental data of Kavasnak et al. (1993), the numerical simulation results of Tian et al. (2007) and Shams et al. (2000), and the empirical equations of Friedlander et al. (1957), Wood (1981) and Fan and Ahmadi (1993) are plotted in this figure together with the proposed model. The experimental data and the simulation results show that the deposition velocities have a "V-shaped" variation. The empirical equation of Friedlander predicts

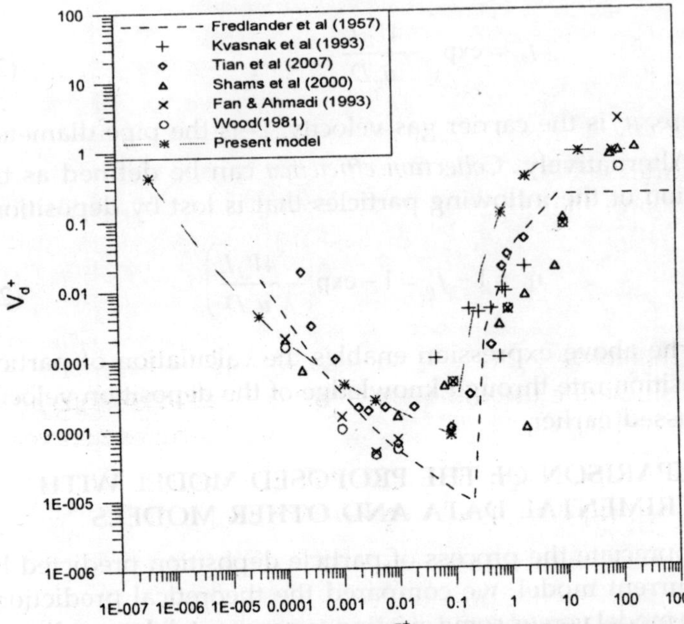

Figure 4: Comparison of Non-dimensional Deposition Velocity as Predicted by Current Model and Earlier Results in Horizontal Pipe

similar trend however the discrepancy with the experimental and numerical results is significant. The models of Wood (1981) and Fan and Ahmadi (1993) are slightly varied for tiny particles. In turn, the suggested model gives better trend and reads the experimental and numerical results quit well.

Three regions can be recognized: the Brownian dominated region, the inertia dominated region, and the very high inertia region. In the Brownian region, the deposition rate increases with decrease in particle size. For large particles, turbulence eddy impaction becomes significant and dominates the deposition process (Chen and Ahmadi, 1997).

For very large particles, the particle inertia becomes very large and essentially not influenced by other forces since the particle deposition rate approaches constant. The tendency of such particles occurs at relaxation time $t^+ > 10$ as shown in Fig. (4).

The deposition rate is calculated using Eq. (27) and compared for all data presented in Fig. (4). Figure (5) illustrates the comparison and one can observe that the current model describes the experimental and numerical data better than other models, in particular, the interface region. Apparently, the inclusion of convective term (flow Reynolds number) in the new model enhances the prediction of particle deposition process specifically in the Brownian regime.

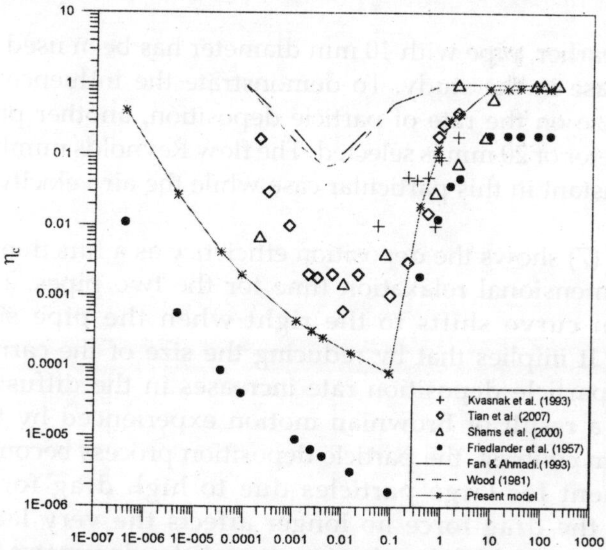

Figure 5: Comparison of Deposition Efficiency Evaluated from Current Model with the Earlier Results in Horizontal Pipe

RESULTS AND DISCUSSION

In this section, influences of particle density, pipe size, and pipe roughness using the proposed model are discussed. Comparison with previous literatures to assess the prediction of particle deposition in a turbulent pipe flow is provided as well.

Particle Density

The effect of particle to fluid density ratio is presented here. Density ratios of 1000 and 2000 are compared to understand the behaviour of the deposition rate responding to particle density.

Figure (6) shows the deposition efficiency as a function of relaxation time for two density ratios. The deposition rate shifts slightly to left with the decrease of density ratio. This trend is in good agreement with that predicted by Chen and Ahmadi (1997) and Fan and Ahmadi (1993).

Pipe Size

As stated earlier, pipe with 40 mm diameter has been used as the base case in the study. To demonstrate the influence of the pipe size on the rate of particle deposition, another pipe with diameter of 20 mm is selected. The flow Reynolds number is kept constant in this particular case while the air velocity is varied.

Figure (7) shows the deposition efficiency as a function of the nondimensional relaxation time for the two pipes. The deposition curve shifts to the right when the pipe size decreases. It implies that by reducing the size of the carrier pipe, the particle deposition rate increases in the diffusion region as a result of Brownian motion experienced by the particles. In contrast, the particle deposition process becomes less efficient for large particles due to high drag force; however, the drag force no longer affects the very large particles since the high inertia of such particles dominates the process of deposition.

Due to scarce in literatures concern with the effect of pipe size, we compared the results of the model with that predicted by Fan and Ahmadi's correlation. Both results are plotted in Figure (8) so one can easily track the behaviour of each model. The only discrepancy between the models is that lie in the impaction region. The conclusion obtained from the correlation of Fan and Ahmadi reads that the deposition rate becomes large for small pipe size in all regions. This could be returned to the fact that particles have tendency to reaches the wall of small pipes faster than that of large pipes. Such conclusion might be accepted only if the carrier velocity for both pipes is same; however in such case Reynolds number

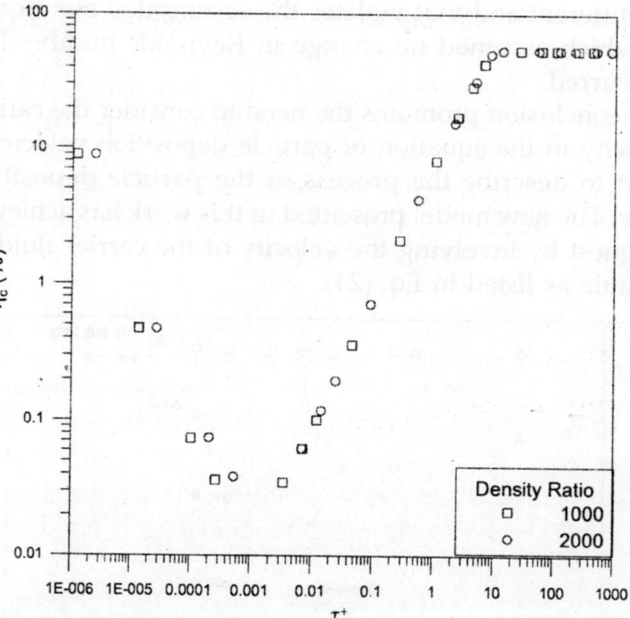

Figure 6: Particle Deposition Efficiency Versus Nondimensional Particle Relaxation Time for Different Density Ratio

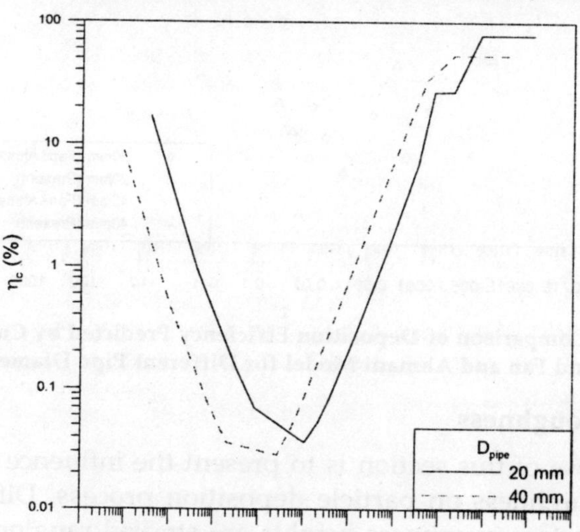

Figure 7: Influence of Pipe Diameter on the Particle Deposition Efficiency

will be different and that violates the assumption mentioned earlier which assumed no change in Reynolds number has been occurred.

This conclusion promotes the need to consider the carrier gas velocity in the equation of particle deposition velocity if we want to describe the process of the particle deposition precisely. The new model presented in this work has achieved such request by involving the velocity of the carrier fluid in the formula as listed in Eq. (24).

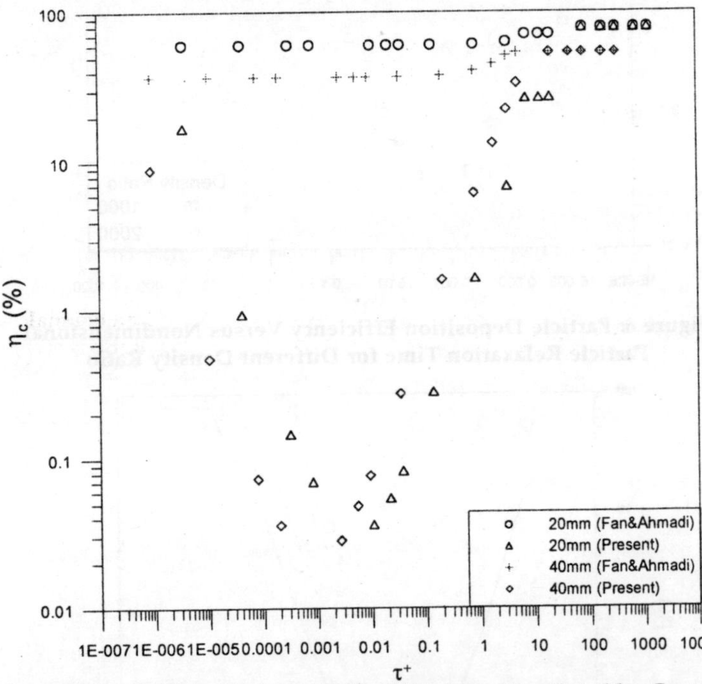

Figure 8: Comparison of Deposition Efficiency Predicted by Current Model and Fan and Ahmadi Model for Different Pipe Diameters

Surface Roughness

The purpose of this section is to present the influence of the surface roughness on particle deposition process. Different average surface roughness heights are studied ranging from smooth surface (k=0) to relatively coarse surface (k~1% of pipe diameter). It is noticed that surface roughness significantly

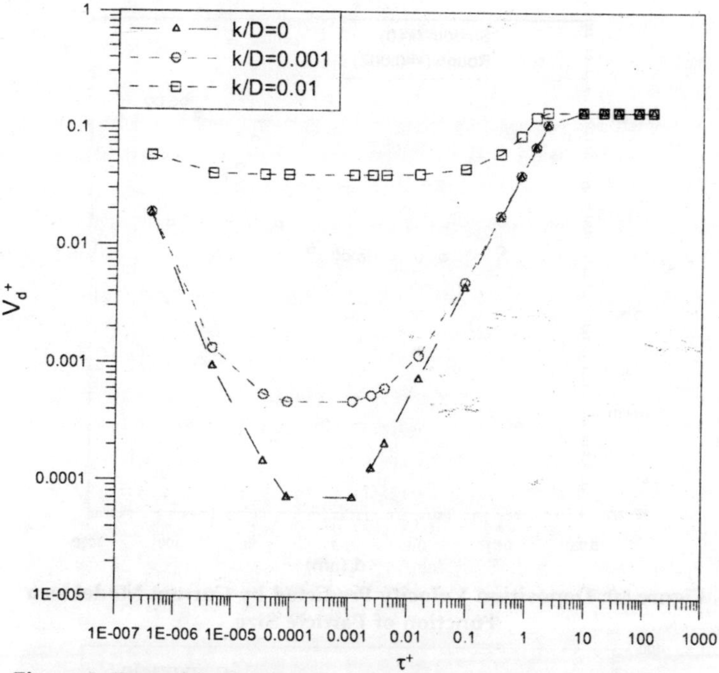

Figure 9: Deposition Velocity as a Function of Relaxation Time at Various Surface Roughnesses

alters the deposition velocity curve particularly in the range of interface region. However, the range encompasses other regions for coarse roughness, as shown in Fig. (9).

Browne (1973) came up with almost congruent conclusion. He studied the deposition velocity for particles ranging from 0.01 to 10 μm flows in a 1.5 cm diameter pipe having relative surface roughness (k/D) of 0.002 and the carrier fluid was air. His conclusion on the effect of surface roughness showed that very large effect occurred as a result of roughness except on particles of sizes larger than ~ 10 microns. Identical conditions are applied to our model to compare the prediction of the deposition velocity using the current expression. Figure (10) illustrates the influence of surface roughness when the modified model is used and one can conclude that the effect of surface roughness decays after ~ 10 μm which is in very good agreement with the Browne's conclusion.

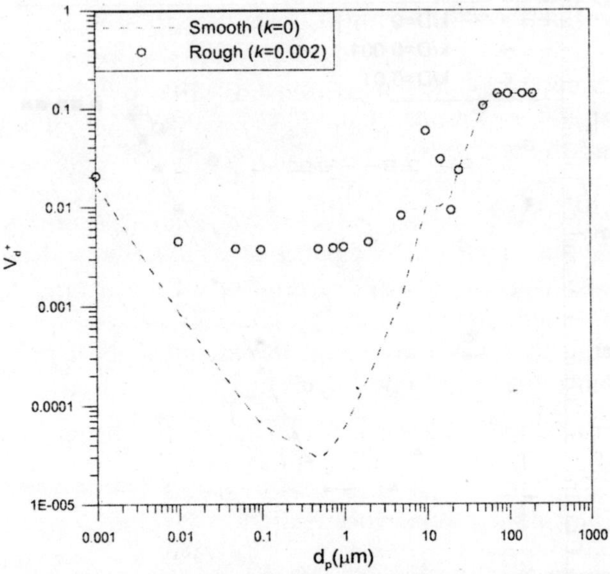

Figure 10: Deposition Velocity Predicted by Current Model as a Function of Particle Size

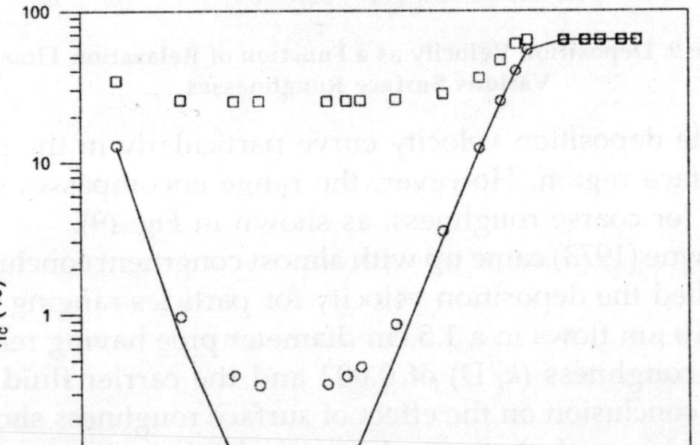

Figure 11: Deposition Efficiency as a Function of Relaxation Time Predicted by Current Model at Different Pipe Roughness

The particle deposition rate is remarkably varied with surface roughness. Figure (11) presents the variation of deposition efficiency as a function of dimensionless relaxation time for different relative roughness. It is clearly shown that the more the coarse pipe wall, the higher the deposition rate.

CONCLUSION

In this work, deposition of particles in a turbulent pipe flow is studied using modified correlation of particle deposition velocity. The influence of the gas velocity has been taken into account in the new model and the analysis showed that the proposed model is in good agreement with the experimental data and numerical results. The deposition efficiency of particles of different sizes, which may be inferred as a measurement of deposition rate, are evaluated from the proposed model and compared with other models available in literatures. The comparison has clearly proved that flow convective process should be included in the particle deposition expression.

The effect of density ratio, carrier size and surface roughness are also studied. Based on the presented results, the following conclusions may be derived:

- Deposition of particles is strongly influenced by local aerodynamics effects.
- The deposition velocity and deposition efficiency in turbulent pipe flow follow V-shaped curve.
- Increase in density ratio results in increase in the deposition rate for submicron sizes.
- Reducing the size of the carrier pipe shifts the particle deposition rate curve to the right. This implies that for particles controlled by diffusion motion, the rate of particles deposit on the pipe wall increases as pipe size decreases. In turn, the deposition rate decreases when the size of pipe decreases in the inertia region.
- Unlike other semi-empirical models, the current correlation describes the influence of surface roughness in a reasonable aspect and its behaviour is in excellent agreement with the previous studies.

- The deposition rate of particles increases in remarkable manner as a result of roughness.

REFERENCES

Browne, L. W. B. Deposition of Particles on Rough Surfaces During Turbulent Gas-flow in a Pipe. *Atmospheric Environment*, 8 (1973), 801-816.

Chen, Q. and Ahmadi, G. Deposition of Particles in a Turbulent Pipe Flow. *Journal of Aerosol Science*, 28 (1997), 789-796.

Cleaver, J. W. and Yates, B. A Sublayer Model for the Deposition of Particles From a Turbulent Flow. *Chem. Engng Sci.* 30 (1975), 983-992.

Crowe, C.T. (2006). Multiphase Flow Handbook. Taylor & Francis Group, FL.

Davies, C. N. (1966). *Aerosol Science*. Academic Press, London.

Fan, F-G. and Ahmadi, G. A Sublayer Model for Turbulent Deposition of Particles in Vertical Ducts With Smooth and Rough Surfaces. *Journal of Aerosol Science*, 24 (1993), 45-64.

Fan, F.-G. and Ahmadi, G. On The Sublayer Model for Turbulent Deposition of Aerosol Particles in The Presence of Gravity and Electric Fields. *Aerosol Science Technology*, 21 (1994), 49-71.

Fichman, M., Gutfinger, C. and Pnueli, D. A Model For Turbulent Deposition of Aerosols. *Journal of Aerosol Science*, 19 (1988), 123-136.

Friedlander, S. K. and Johnstone, H. H. Deposition of suspended particles from turbulent gas stream. *Industrial and Engineering Chemistry*, 49 (1957), 1151-1156.

Fuchs, N. A. (1964). *The Mechanics of Aerosol*. Pergamon, Oxford.

Hidy, G. M. (1984). *Aerosols, an Industrial and Environmental Science*. Academic Press, New York.

Hinds, W.C., (1999). *Aerosol Technology: Properties, Behaviour, and Measurement of Airborne Particles*. 2nd Edition, Wiley, New York.

Hussein T., Hruška A., Dohányosová P., Dzumbová L., Kulmala M. Deposition Rates on Smooth Surfaces and Coagulation of Aerosol Particles Inside a Test Chamber, *Atmospheric Environment*, 43(4) (2009), 905-914.

Ilori, T. A. (1971). Turbulent Deposition of Aerosol Particles Inside Pipes. PhD. thesis, University of Minnesota.

Kvasnak, W., Ahmadi, G., Bayer, R., & Gaynes, M. Experimental Investigation of Dust Particle Deposition in a Turbulent Channel Flow. *Journal of Aerosol Science*, 25 (1993), 795-815.

Li, A. and Ahmadi, G. Dispersion and Deposition of Spherical Particles from Point Sources in a Turbulent Channel Flow. *Journal of Aerosol Science Technology*, 16 (1991), 209-226.

Li, A. and Ahmadi, G. Deposition of Aerosols on Surfaces in a Turbulent Channel Flow. *International Journal of Engineering Science*, 31 (1993), 435-451.

Li, A., Ahmadi, G., Bayer, R. G. and Gaynes, M. A. Aerosol Particle Deposition in an Obstructed Turbulent Duct Flow. *Journal of Aerosol Science*, 25 (1994), 91-112.

Liu, B. Y. H. and Agarwal, J. K. Experimental Observation of Aerosol Deposition in Turbulent Flows. *Journal of Aerosol Science*, *5* (1974), 145-155.

Maxwell J. C., On the Dynamical Theory of Gases. *Philos. Mag.*, 2 (1868), 26-78.

Miguel, A. F., A. F., Reis, A., & Aydin, M. Aerosol Particle Deposition and Distribution in Bifurcating Ventilation Ducts. *Journal of Hazardous Materials*, B116 (2004), 249-255.

Papavergos, P. G. and Hedley, A. B. Particle Deposition Behavior from Turbulent Flow. *Chemical Engineering Research and Design*, 62 (1984), 275-295.

Rashidi, M., Hetsroni, G., Banerjee, S. Particle-Turbulence Interaction in a Boundary Layer. *International Journal of Multiphase Flow*, 16 (1990), 935.

Schwendiman, L. C. and Postma, A. K. (1961). Hanford Laboratory Report HW-SA-2236, Richland, Washington, U.S.A., also (1962) Tech. Inf. Div. Report TID-7628, Book 1 (USAEC) 1.

Sehmel, G. A. (1968). Aerosol deposition from turbulent airstreams in vertical conduits, Batelle Northwest Laboratory Report BNWL-578, Richland, Washington, U.S.A.

Shams, M., Ahmadi, G., & Rahimzadah, H. A Sublayer Model for Deposition of Nano- and Micro-particles in Turbulent Flows. *Chemical Engineering Science*, 55 (2000), 6097-6107.

Sosnowski, T. R., Moskal, A., & Gradon, L. Mechanism of Aerosol Particle Deposition in the Oro-pharynx Under Non-steady Airflow. *Annals of Occupational Hygiene*, 51 (2007), 19-25.

Tian, L. and Ahmadi, G. Particle Deposition in Turbulent Duct Flow-comparisons of Different Model Predictions. *Journal of Aerosol Science*, 38 (2007), 377-397.

Wells, A.C. and Chamberlain, A.C. Transport of Small Particles to Vertical Surfaces. *British Journal of Applied Physics*, 18 (1967), 1793.

Wells, A. C. and Chamberlain, A. C. Deposition of Dust From Turbulent Gas Streams. *Atmospheric Environment*, 3 (1969), 494-496.

Wood, N. B. A Simple Method for the Calculation of Turbulent Deposition to Smooth and Rough Surfaces. *Journal of Aerosol Science*, 12 (1981), 275-290.

10

ISOLATION AND CHARACTERIZATION OF NICKEL-RESISTANT BACTERIA FROM ELECTROPLATING WASTE

Susmita Mishra, Alok Prasad Das and Padma Seragadam

INTRODUCTION

Pollution due to heavy metal toxicity is an ever mounting problem in the developing countries. Heavy metals are key pollutants in marine, ground, industrial and even treated wastewater Valdman *et al.*, (2001). Toxic heavy metals are found in effluents and discharged wastewater of industries like electroplating, steel, alloy, motor vehicles, aircraft, paint, chemical, textile, pigment etc. (Saithong etal., Prasertsan *et al.*, 2002). Release of Ni, Cd, Cr & Pb into the environment has increased in the recent years at an alarming rate. Nickel is well known heavy metal pollutant, which is present in effluents of electroplating industries to the tune of 20-200 ppm (Revathi *et al.*, 2005). Studies revealed that excessive exposure to nickel may effects lungs, nose, kidney, reproductive system, skin and fertility. In prolonged and direct contact with skin, nickel may cause an allergic reaction on nickel sensitized people. Chronic inhalation and exposure to nickel fumes may cause cancer. The acceptable limit of Ni in drinking water and industrial discharge is limited to 0.01 mg/L and 2 mg/L respectively (Sharma *et al.*, 1992).

Toxicity of nickel to aquatic organisms such as bacteria (Giashuddin *et al.*, 1979), invertebrates (Biesinger *et al.*, 1972) and fish (Pickering., 1974) is well-known. So it is very essential to eliminate Ni from industrial effluents. Some of the conventional processes which are used for the exclusion of nickel from the industrial effluents are chemical precipitation, oxidation or reduction, electrochemical treatment, filtration, ion exchange, and membrane technologies. These methods may be unproductive or expensive, particularly when the heavy metal ions in the contaminated media are high i.e. in order of 1-100 mg dissolved heavy metal ions/l (Volesky *et al.*, 1990). Secondly the operational troubles and the high cost of handling necessitate the research for some newer methods. Biological methods such as biosorption/ bioaccumulation may provide an attractive alternative to conventional methods for the removal of heavy metal ions (Kapoor *et al.*, 1995).

Among the heavy metal-resistant organisms, the nickel-resistant bacteria offer the unique opportunity to compare bacteria isolated from anthropogenically nickel-polluted ecosystems with those isolated from naturally nickel-percolated ecosystems. The state-of-the-art of bacterial transport systems for the controlled uptake of nickel was recently reviewed by (Eitinger *et al.*, 2000). A nickel resistant, hyper-accumulating strain of the fungus *Neurospora crassa* was described by (Kumar *et al.*,1992), but there is no information about its use in bioremediation. Congeevaram *et al.*, (2007) could achieve 55% removal of Ni (II) from its aqueous solution using a *micrococcus sp.*

In the present study we have tried to isolate yet another new species from the electroplating industry waste water responsible for Ni (II) degradation. To explore the potential of the selected species for treatment of Nickel waste water influence of various parameters such as initial pH, contact time, initial nickel concentrations etc on nickel degradation has been investigated.

MATERIALS AND METHODS

Isolation of Microorganisms

For isolation and enumeration of microorganisms, samples obtained from metal contaminated soil were serially diluted in sterile distilled water and plated on Nutrient agar media. Plates were incubated at 30 ±2 °C respectively for 2-4 days and colonies differing in morphological peculiarities were isolated in pure form and maintained on slopes of respective agar.

Screening for Nickel-Resistance

Screening of Nickel resistant organism helps in selecting the organism responsible for nickel degradation. MIC is defined as the lowest concentration of metal at which no colony forming units (CFU) are observed. Minimum inhibitory concentration (MIC) was tested with varying nickel concentrations from 50 mg/L to 5000 mg/l.

Selection of Potent Strains

Based on the growth performance in nickel-containing Nutrient agar, nickel-resistant strains were selected. The isolate was identified following morphological and

Table 1
Morphological and Biochemical Characteristics of Nickel Resistant Organism

	Characteristics	*Response*
1. Morphological	Gram stanning	+
	Motility	+
	Colony Morphology	Coccus in rod shape
	Colour	Creamish White
2. Biochemical	Indole Production	–Ve
	Citrate utilizing	–Ve
	VP reaction	–Ve
	MR reaction	+Ve
	Oxidase activity	–Ve
	H_2S production	–Ve
	Starch hydrolysis	–Ve

biochemical characterization according to Bergey's Manual of Systematic Bacteriology (Holt *et al.*, 1989) Table 1.

Characterization of Electroplating Waste Water

Electroplating effluent samples from two different sites of a Steel plant (after precipitation unit), Orissa, India were collected in sterile containers and brought to the laboratory. The samples were characterized based on their metal contents and the average values are displayed in Table 2.

Table2

Concentrations of Heavy Metals Before & After Treatment

Metals	*Initial Concentration (mg/L)*	*Final Concentrations (mg/L) after treatment (24hours)*
Copper (Cu)	3.25	Below detection limit
Chromium (Cr)	50	6.02
Nickel (Ni)	50	4.5
Zinc (Zn)	250	7.9

Table 3

Comparison of Nickel Degradation Characteristics by Different Species

Metal	*Microorganism*	*Initial Concen-tration (mg/L)*	*pH*	*% re-moval*	*Time (hrs)*	*Reference*
NI(II)	Malasian Rubber	20	5	65	4	S. Hasan *et al.*
NI(II)	Lemina Minor	5	7-9	87	5	R. A. Nicholas *et al.*
NI(II)	Micrococcus	50	7	55	20	S. Congeevaram *et al.*
NI(II)	Aspergillus Niger	50	5	90	20	Congeevaram *et al.*
Ni(II)	Staphylococcus sp.	50	7	97	12	Present study

Measurement of the Kinetics of Cellular Growth and Heavy Metal Removal

Bacterial isolate was added into a 250 ml flask containing 100 mg L^{-1} of Ni (II).The flask was kept in a incubator shaker (150 rpm) at solution pH and fixed temperature of 28°C for 24 h. During the incubation period, Ni (II) concentration and biomass were monitored for every 4 hours interval until it attains a saturation level. To study the influence of initial pH of the nickel solution 1M HCl and 1M NaOH was used to adjust the pH from 5-8. To explore the tolerance of the isolates to Ni (II), optimal culture conditions were used with varying initial Ni (II) concentrations (25, 50, 100, 200 & 400 mg/L). All the cultures including controls were incubated. After every 4 hrs regular interval, the cell growth was measured. The extent of tolerance was compared i.e., biomass at each Ni (II) concentration per biomass using a control. To ensure accuracy of the results each experiment were conducted in duplicate.

Heavy Metal Assay and Biomass Quantification

Ni (II) concentration was estimated using di-methylglyoxime (DMG) at its maximum wavelength 600 nm. Estimation of Ni (II) using DMG is a very sensitive method and it can estimate as low as 1/1000 mg. Bacterial biomass was quantified using spectrophotometer (JASCO UV/Vis-550) at l_{max} 595 nm (Congreevan *et al.*, 2007). The initial and the final concentration of heavy metals used in batch mode studies were calculated by estimating the concentration of metals spectrophotometrically. From the difference in concentration the removal efficiencies of the microorganism has been calculated (W. Armit *et al.*, 1906).

Resistance to Other Heavy Metals

The waste water from electroplating industry contains other heavy metals, hence it is essential to study the resistively of the selected organism to different heavy metals. Resistance of selected bacterial isolate to other heavy metals was tested in nutrient agar supplemented with metals. Metals that were tested include $ZnSO_4$, $CoCl_2$, $K_2Cr_2O_7$, $Pb(NO_3)_2$, $CuSO_4$, $MnSO_4$ & $HgCl_2$. All the Ni (II) resistant bacterial strains are screened for their resistance to the salts of these heavy metals. The strains are inoculated on nutrient agar supplemented with 100mg/l concentrations of metallic salts and incubated at 30°C for 24 hrs.

RESULTS AND DISCUSSION

The Concentrations of some of the heavy metals collected from electroplating industry after the precipitation unit are given in Table 2. It was interesting to note that the nickel content of the samples was 50 mg/L. The other heavy metals such as Cu, Zn, and Cr were also estimated and were found to be above their permissible limits.

Screening of Microorganisms for Nickel-Resistance

Sixteen bacterial populations were isolated in pure form from isolation plates and subjected to screening for nickel-resistance in Agar media containing 25-400 mg/L of Ni^{2+}. The majority of the isolates were able to grow in 50 mg Ni^{2+}/l but their number decreased as the metal concentration in the media was increased. *Staphylococcus aureus* was highly resistant to nickel and showed visible growth above 200 mg Ni^{2+}/L. These results confirm the nickel-resistance potential of bacterial

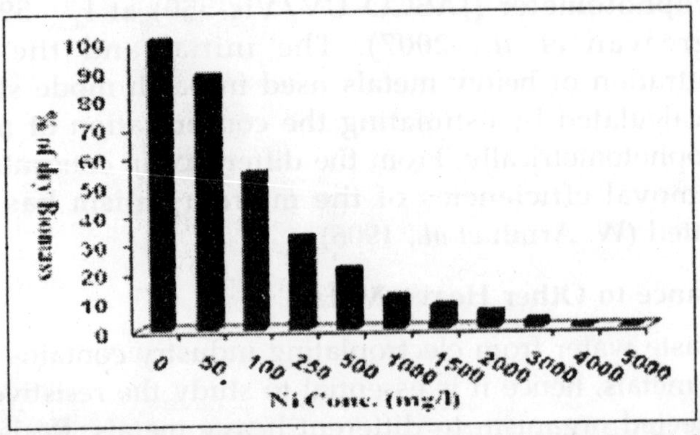

Figure 1: Minimum Inhibitory concentration of Ni on microbial growth

strains. Minimum Inhibitory Concentration (MIC) for Nickel is found to be 4000-5000mg/l as above this concentration there was no visible growth of the bacteria Figure 1. Similarly Congreevan et al in 2007 reported that MIC for Ni (II) by Micrococcus & Aspergillus sp. as 500 mg/l.

Influence of Solution pH on Nickel Degradation

The initial Nickel (II) pH range is studied from 5-8 in the interval of 0.5 . The initial nickel concentration in the aqueous solution was maintained at 100 mg/L with varying initial pH of the solution. The maximum removal of Ni (II) is observed at the neutral pH 7 Figure 2. Shankar Congeevaram *et al.*, (2007) studied the biosorption of Nickel using micrococcus bacteria at optimum pH 7 and the initial concentration 50 mg/l. The decrease in removal of Ni (II) above pH 7 is due to the formation of $Ni(OH)_2$. Substantial precipitation of nickel as nickel hydroxide occurs at high pH values. The formation of hydroxide precipitate reduces the amount of free nickel ions, which accumulates to the organism. Similar reports has been reported for Ni (II) removal using *S. cerevisiae*, where the outer cell wall consists of protein coat, which develops a charge by the dissociation of ionizable side groups of the constituent amino acids. The ionic state of ligands such as carboxyl, phosphate, imidazole, and amino groups will promote reactions with the positively charged metal ions. At low pH, cell walls ligands were closely associated with the hydronium ions [$H3O^+$] and restricted the approach of metal cations as a result of the repulsive forces. The readings were further confirmed from biomass growth which was maximum at pH 7.

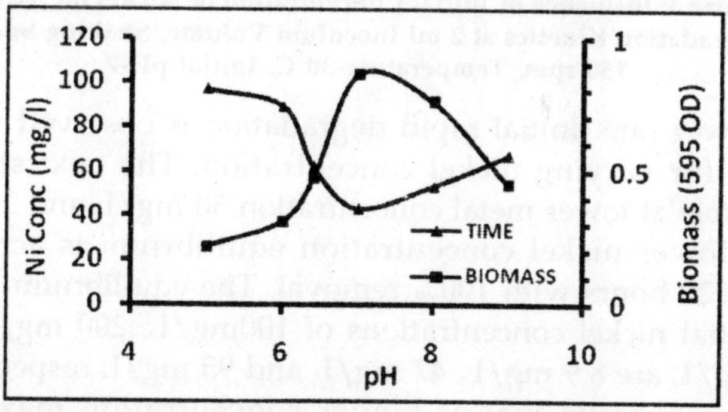

Figure 2: Influence of Initial pH of the Nickel Solution on Nickel Degradation at 2 ml Innoculum Volume, Shaking Speed 150 rpm, Temperature-30°C, Initial Nickel Concentration-50 mg/L, Equilibrium Time-24 hours

Measurement of the Kinetics of Broth Cellular Growth and Heavy Metal Removal at Different Intial Concentrations

The Ni (II) reduction ability of the bacteria was growth dependent & the *Staphylococcus* sp. reduces nearly 400 mg/L of Ni (II) during degradation experiment in 60-h Figure 3. It is evident from the experiment that the time required for Ni (II) degradation varies with initial nickel concentration. The time required for nickel degradation increased with initial nickel concentration. But the rate of nickel removal showed little change at 200 mg/L nickel concentration. We can also find three different stages of Nickel degradation with initial rapid stage followed by slow rate and finally degrades at a

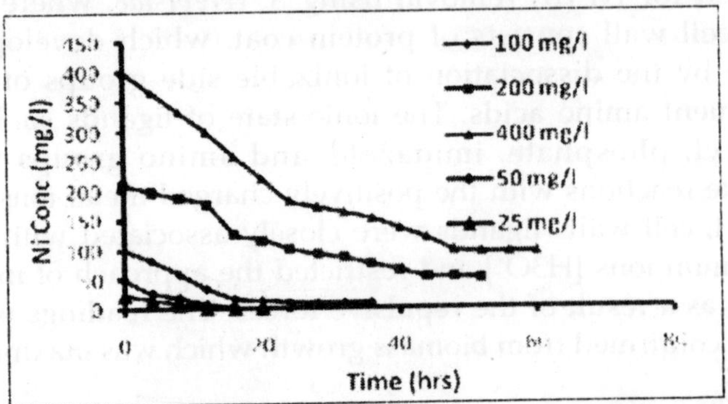

Figure 3: Influence of Initial Concentration of Nickel on Nickel Degradation Kinetics at 2 ml Inoculum Volume, Shaking Speed 150 rpm, Temperature-30°C, Initial pH-7

till slower rate. Initial rapid degradation is observed within 20 hrs for varying nickel concentration. The next stage is negligible at lower metal concentration 50 mg/L and 25 mg/L. At lower nickel concentration equilibrium is achieved within 20 hours with 100% removal. The equilibrium values for initial nickel concentrations of 100mg/L, 200 mg/L and 400 mg/L are 8.9 mg/L, 47 mg/L and 95 mg/L respectively. We can observe that at higher concentration maximum reduction is observed within 20 hrs at initial nickel concentration 400 mg/L than 200 mg/L. This is attributed to high concentration gradient of nickel in the solution which is the main driving force for nickel degradation. Various researchers have used the optimum conditions for nickel degradation using different species as represented in Table 2.It reflectes that *staphylococcus aureus* has maximum potential to degrade nickel in minimum time.

The innoculum of the bacterial strains cultured overnight was used for this experiment. Culture flasks (150 ml) with a final volume of 100ml supplemented with (25-400 mg/L) of Ni were inoculated with 2ml of inoculums for 24 hour. The growth kinetics of bacteria is characterized as initial lag phase, second exponential phase, stationary phase and death phase. In this experiment it is observed that lag phase is increasing with increased initial Ni concentration Figure 4. It is basically due to inhibitory effect of higher Ni concentration on the growth of the organism. Each organism has a specific resistance at a specified growth condition. As the initial age of the innoculum was fixed at 24 hours the acclimatization period at varying Ni concentration will not remain same. Hence the following behavior is observed. The Ni-resistant bacteria isolate exhibited reduced bioaccumulation when cells were in stationary phase. At higher concentrations the growth of

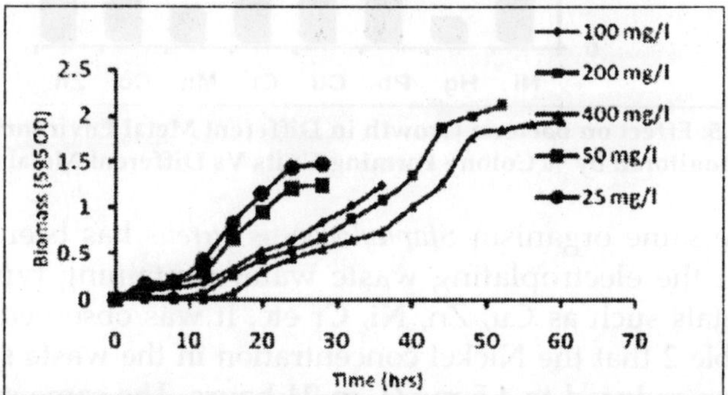

Figure 3: Influence of Initial Concentration of Nickel on Nickel Degradation Kinetics at 2 ml Inoculum Volume, Shaking Speed 150 rpm, Temperature-30°C, Initial pH-7

the bacteria is inhibited due to fixed amount of innoculum for all the different concentration of Ni considered in the experiment. (G. Cabrera *et al.*, 2005)

Resistance to Other Heavy Metals

The selected nickel-resistant strains were tested for their resistance to other heavy metals such as Zn^{2+}, Pb^{2+}, Cu^{2+}, Cd^{2+}, Co^{2+}, Cr^{6+}, Mn^{2+}. Relative growth of isolate in different metal-containing media is shown in Figure 5. It was evident that isolate was resistant to all metals at low concentration but the metals appear to be toxic at higher concentration. The promising isolate was tested for its tolerance to other heavy metals. The isolated Sp. showed high degree of resistance to different heavy metals in the order Zn > Ni > Cr > Cu > Pb > Co > Mn > Hg. It indicates that the bacterial sp. can be used to degrade the following metals at their optimized conditions.

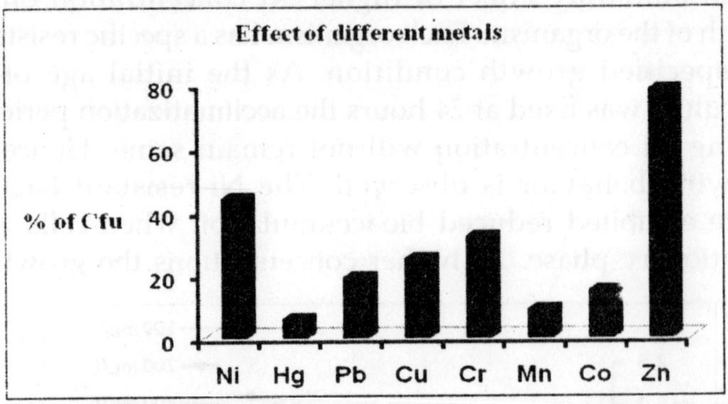

Figure 5: Effect on Bacteria Growth in Different Metal Environments, Monitored by % Colony Forming Units Vs Different Metals

The same organism *Staphylococcus aureus* has been used to treat the electroplating waste water containing typically the metals such as Cu, Zn, Ni, Cr etc. It was observed from the Table 2 that the Nickel concentration in the waste stream has been reduced to 4.5 mg/L in 24 hours. The same species showed 97% removal of Nickel in its aqueous solution within 12 hours. It can be ascribed due to the presence of other heavy metals in the effluent that inhibited the growth of organism.

CONCLUSION

These studies clearly indicated that the electroplating waste containing nickel along with other heavy metals can be removed at its lower concentration effectively using the identified species Staphylococcus *aureus*. This bacterium effectively removed the Ni (ii) upto 97% in 12 hrs at its optimum condition such as neutral pH, with inoculums size 2 ml/100 ml of solution. Its MIC for Ni (II) is estimated as 5000 mg/L. The isolated Sp. showed high degree of resistance to different heavy metals in the order Zn > Ni > Cr > Cu > Pb > Co > Mn > Hg. This needs further research as it may be proved to be cheapest method of treatment of electroplating wastes.

REFERENCE

Valdman E., Erijman L., Pessoa F. L. P., and Leite S. G. F. (2001), Continuous Biosorption of Cu and Zn by Immobilized Waste Biomass Sargassum sp., *Process Biochemistry*, 36: 869-873.

Saithong K., and Prasertsan P. (2002), Biosorption of Heavy Metals by Thermo Tolerant Polymer Producing Bacterial Cells and the Bioflocculant, *Sangklanakarin Journal of Science and Technology*, 24: 421-430.

Eitinger T. and Mandrand-Berthelot M. A. (2000), Nickel Transport Systems in Microorganisms, *Arch Microbiol*, 173: 1-9.

Kumar S. C., Sastry S. K. and Mohan M. P. (1992), Use of Wild Type and Nickel Resistant *Neurospora Crassa* for Removal of Ni^{+2} from Aqueous Medium, *Biotechnol Lett*, 14: 1099-1102.

Robinson B. H., Chiarucci A., and Brooks R. R. (1997), The Nickel Hyperaccumulator Plant *Alyssum bertolonii* as a Potential Agent for Phytoremediation and Phytomining of Nickel, *J Geochem Explor*, 59: 75-86.

G. Cabrera, J. M. Gómez, and D. Cantero (2005), Kinetic Study of Ferrous Sulphate Oxidation of *Acidithiobacillus Ferrooxidans* in the Presence of Heavy Metal Ions, *Enzyme and Microbial Technology*, 36: 301-306.

Revathi M. (2005), Removal of Nickel Ions from Industrial Plating Effluents Using Activated Alumina as Adsorbent, *J. of Environmental & Engg*, 47: 1.

Volesky B. (1990), Biosorption and Biosorbents, In: Biosorption of Heavy Metals, Boston, *USA, CRC Press*, 3-5.

Kapoor A., and Viraraghavan T. (1995), Fungal Biosorption, An Alternative Treatment Option for Heavy Metal Bearing Wastewater: A Review, *Bioresource Technolology*, 53(3): 195-206.

Sharma Y. C., Prasad G., Rupainwar D. C. (1992), Removal of Ni (II) from Aqueous Solutions by Sorption, *Interl J. of Env. Stud*, 40: 535.

Mance G. (1987), Pollution Threats of Heavy Metals in Aquatic Environments, *Elsevier Applied Science*, New York.

Giashuddin M., Cornfield A. H. (1979), Incubation Study on Effects of Adding Varying Levels of Nickel (As Sulphate) on Nitrogen and Carbon Mineralization in Soil, *Environ Pollut*, 15: 231-234.

Biesinger K. E., and Christensen G. M. (1972), Effects of Various Metals on Survival, Growth, Reproduction, and Metabolism of Daphnia Maqna, *J Fish Res Bd Can*, 29: 1691-1700.

Pickering Q. H. (1974), Chronic Toxicity of Nickel to the Fathead Minnow, *J. Water Pollut Control Fed*, 46: 760-765.

Holt J. G., Sneath P. H. A., Mair N. S., and Sharpe M. E. (1989), Bergey's Manual of Systematic Bacteriology, 2: 1104-1139.

Rubikas J., Matulis D., Leipus A., and Urbaitiene D. (1997), Nickel Resistance in Escherichia Coli V38 is Dependent on the Concentration Used for Induction, *FEMS Microbiology Letters*, 155: 193-198.

Armit W., and Harden A. (2007), The Quantitative Estimation of Small Quantities of Nickel in Organic Substances, Proceedings of the Royal Society of London, Series B, *Containg Papers of a Biological Character*, 77: 420-423.

Congeevaram Shankar, Dhanarani Sridevi, Park Joonhon., Dexilin Michael, and Thamaraiselvi Kaliannan, (2007), Biosorption of Chromium and Nickel by Heavy Metal Resistant Fungal and Bacterial Isolates, *J. Hazardous Materials*, 146: 270-277.

S. Hasan, Mohd Ali Hasim, and B. S. Gupta (2000), Adsorption of Ni (SO_4) on Malaysian Rubber-Wood *Ash. Biores., Technology*, 72: 153-158.

R. A. Nicholas, S. P. K. Stenberg, and C. Kathryn (2003), Lead Nickel Removal Using Microspora and Lemina Minor, *Biresource Technology*, 89: 41-48.

11

AQUIFER PARAMETER IDENTIFICATION WITH SIMULATED ANNEALING ALGORITHM AND ITS APPLICATION

Shouju LI, Yingxi LIU, He YU

INTRODUCTION

Distributed parameter models are used to simulate groundwater flow, but the reliability of results is strongly dependent on whether the model parameters are properly identified. However, the real geological structure of an aquifer is generally very complex and mostly unknown. More data (Tan,2008) are required for identifying a more complex parameter structure, and there is always the problem of over-parameterization. Additionally, the inverse problem of parameter structure identification is highly non-linear and non-convex; hence, gradient-based algorithms often get trapped in local optima. On the other hand, heuristic search algorithms usually require a very large number of simulation runs. There remains a need to develop efficient inverse solution algorithms. A common use of groundwater flow models (Carrera, 1986) is to predict the response of an aquifer.

While the mathematical and computational aspects of such response predictions are reasonably well developed, the question of how to choose appropriate parameter values for a specific aquifer has not been completely resolved. Traditionally, the determination of aquifer parameters (Yeh, 1986) is based on trial-and-error and graphical matching techniques under the assumptions that the aquifer is homogenous and isotropic and a closed-form solution for the governing equation exists. The inverse problem of aquifer parameter identification is often ill-posed. The ill-posedness is generally characterized by the nonuniqueness and instability of the identified parameters. The instability of the inverse solution stems the fact that small errors in heads will cause serious errors in the identified parameters. Classical identification procedures (Zheng, 1996) are based on the optimization algorithm. Their drawbacks lie in lacking robustness and global convergence property. Parameter identification, or model calibration, is a critical step in the application of mathematical models in hydrologic sciences. Unfortunately, parameter identification is an inherently difficult process and, as an inverse problem, it is plagued by the well-documented problems of nonumiquenes, nonidentifiability and instability.

Numerous optimization techniques (Kool, 1987) have been used to solve groundwater remediation design and parameter identification problems. In a parameter identification problem, the objective function can be the weighted difference between the observed and calculated values at certain observation points in the aquifer. The identified parameters can be hydraulic conductivity or other aquifer parameters, such as storage coefficient. Zadeh (2008) proposed a new strategy which coupled a mass-lumped Galerkin linear finite element solution of the mixed form Richards equation with an experimental time-space series and the Osborne–Moré revised form of the Levenberg–Marquardt algorithm; to retrieve hydraulic parameters of a partially saturated porous medium. The numerical simulator showed excellent agreement with a reference solution, obtained on a dense grid and infinitesimal time step, in terms of fluid pressure head, fluid content, and fluid volumetric flux density and perfectly conserves the global mass. An adaptive algorithm was implemented to estimate sensitivity matrix in the inverse algorithm. A multi-criterion stopping rule was developed and successfully implemented to end the inverse code at the solution. Rivest (2008) presented an approach to hydraulic head mapping based on kriging with an external drift in which the auxiliary variables were obtained by finite-element modeling. The approach relied on the idea that numerical solutions stemming from simple conceptual models were suitable candidates for the hydraulic head field drift. Indeed, these numerical head fields obeyed the groundwater flow equations and incorporate information gathered during site investigations such as parameter estimates, geology and boundary conditions. Eldho(1997) developed the dual reciprocity boundary element method for the simulation of contaminant transport problems in porous media. A detailed sensitivity analysis was carried out to find the optimal dual reciprocity boundary element model parameters. Qin (2009) gave a comprehensive review on recent developments, advancements, challenges, and barriers associated with simulation and optimization techniques in supporting process control of petroleum waste management and site remediation. A number of related methodologies and applications were examined. Perspectives of effective site management were investigated, demonstrating many demanding areas for enhanced research efforts, which include issues of data availability and reliability, concerns in uncertainty, necessity of post-modeling analysis, and usefulness of development of process control techniques. To improve the soil moisture estimation, Reshmidevi (2008) proposed a new SCS-CN-based model for soil moisture simulation. The proposed model is built up on the soil moisture balance equation to account for the effect of ponding condition and soil moisture variation between the dry and saturation condition. Young (2006) proposed an innovative computational model for the large eddy simulation modeling of multi-dimensional unsteady turbulent flow problems in external flow field. Based on the LES principles, the model uses a pressure projection method to solve the Navier-Stokes equations in transient condition. The present model is used to simulate the flows past a two-dimensional square rib and a three-dimensional cube at high Reynolds number. The simulation results are found to be reasonable and comparable with other models available in the literature even for coarse meshes.

In recent years, global optimization methods (Wang, 1998) are being increasingly used to solve groundwater remediation design and parameter identification problems. These methods include simulated annealing, genetic algorithm, tabu search and ant colony system. Compared with gradient based local search methods, global optimization methods do not require the objective function to be continuous, convex, or differentiable. They have also shown other attractive features such as robustness, ease of implementation, and the ability to solve many types of highly complex, nonlinear problems. One common drawback of these global optimization methods is that many objective function evaluations are typically required to obtain optimal or near-optimal solutions. Bau (2008) presented a stochastic optimal control framework for assisting the management of the cleanup by pump-and-treat of polluted shallow aquifers. In the problem being investigated, hydraulic conductivity distribution a location were considered as the uncertain variables. The framework considered the subdivision of the cleanup horizon in a number of stress periods over which the pumping policy implemented until that stage was dynamically adjusted based upon new information that has become available in the previous stages. Tan (2008) embed an adjoint state procedure in the search process to improve the efficiency of the Tabu search. Voronoi tessellation for automatic parameterization was used to reduce the dimension of the distributed parameter. Additionally, a coarse-fine grid technique was applied to further improve the effectiveness and efficiency of the proposed methodology. To avoid over-parameterization, at each level of parameter complexity, the residual error was calculated for parameter fitting, the parameter uncertainty error and a modified Akaike Information Criterion. Ayvaz (2007) proposed an inverse solution algorithm through which both the aquifer parameters and the zone structure of these parameters can be determined based on a given set of observations on piezometric heads. In the zone structure identification problem fuzzy c-means (FCM) clustering method was used. The association of the zone structure with the transmissivity distribution was accomplished through an optimization model. The meta-heuristic harmony search (HS) algorithm, which was conceptualized using the musical process of searching for a perfect state of harmony, was used as an optimization technique. The optimum parameter zone structure was identified based on three criteria which are the residual error, parameter uncertainty, and structure discrimination. The simulated annealing algorithm is a kind of natural algorithm inspired by behavior or processes presented in nature. When compared with traditional first-order methods, the simulated annealing algorithm is recognized to have a better capability to find the global optimum solution. The objective of this paper is to present a new method based on simulated annealing algorithm for obtaining the parameters of a linear groundwater flow model.nd dissolved contaminant plume

NUMERICAL SIMULATIONS OF GROUNDWATER FLOW MODELS

Solutions for groundwater seepage problems have been developed since the pioneering work of Henry Darcy (1856). For example, many analytical solutions were developed and presented by Harr (1962) and Polubarinova-Kochina (1962). The groundwater seepage problem can be described by Darcy's law and continuity equations. The seepage equation is obtained by combining these two equations. Since the seepage equations are based on these two equations, the assumptions and limitations that apply to these equations also apply to the seepage equation. Darcy's law basically demonstrates a linear dependency between the hydraulic gradient and the discharge velocity. In the groundwater and seepage problems, where the flow is primarily of a single relativity incomprehensible fluid subject to small changes in temperature, it is more convenient to use Darcy's law with k, coefficient of permeability. Physically, all flow systems extend in three dimensions. However, in many problems the features of the groundwater motions are essentially planar, with the motion being substantially the same in parallel planes. Therefore these problems can be described with two-dimensional flow only, and thereby reduce considerably the work necessary to obtain a solution.

Consider a two dimensional unsteady seepage field, as shown in Fig.1, its governing equation can be written as the follows:

$$k_x \partial^2 h / \partial x^2 + k_y \partial^2 h / \partial y^2 = q + S \frac{\partial h}{\partial t} \tag{1}$$

Where k_x and k_y are conductivities in x and y direction ,respectivelyÿh is the water head, q *is* the source-sink item, S is storage coefficient. The first kind boundary condition is expressed as the following:

$$h(x,y)|_{\Gamma_1} = h_0(x,y) \tag{2}$$

Where $h_0(x,y)$ is the already known head. The second kind boundary condition is written as the following:

$$Q_0(x,y)|_{\Gamma_2} = k_x \frac{\partial h}{\partial x} l_x + k_y \frac{\partial h}{\partial y} l_y \tag{3}$$

Where $Q_0(x,y)$ is the drainage already known, l_x ,and l_y are the direction cosines of the exterior normal of the boundary in x and y direction , respectively. The initial condition is listed as follows

$$h(x,y,t=0) = h_i(x,y) \tag{4}$$

The following finite element equation can be derived by the principle of the variational method:

$$[G]\{h\} = \{F\} \tag{5}$$

Where [G] is the matrix of the water transmissibility coefficient, which is already known, and {F} is the free item. When the boundary condition, initial condition and the model parameters are determined, the infinite element equation is adopted to compute the distribution of the water head and the drainage in the whole seepage field, which provides model data to the analysis of the inversion problem of the aquifer parameter estimation.

AQUIFER PARAMETER IDENTIFICATION WITH SIMULATED ANNEALING ALGORITHM

Solution definition of Inverse Problem and its Ill-posedness in Groundwater Problem

The parameter identification problem can be formulated to find the model parameters by adjusting *m* until the measured data match the corresponding data computed from the parameter set in a least-squares fashion. The objective function (Abbaspour, 2001) is defined as follows

$$J(m) = (h_m - h_c(m))^T W(h_m -_c (m)) \qquad (6)$$

Where h_m is the measured waterhead vector; h_c is the computing waterhead vector, which is related to the identified parameter vector *m*. *w* is weighting matrix in order to take into account the different observed equipments for the water head measurements. This objective function clearly depends on the measured data and the parameters of model. Fig. 1 shows the groundwater flow model, in which the four observing points for water heads are set in order to get measurement data and to identify the aquifer parameters. The objective function can become complex as shown as Fig.2, such as non-convex, or even multi-modal if errors contained in the model equation or /and errors in the measurement data are large. The multi local minima can be found from Fig.2. In such a case, the solution may vibrate or diverge when conventional gradient-based optimization methods are used, which gives rise to the necessary for a robust optimization method such that a stable convergence is always achieved.

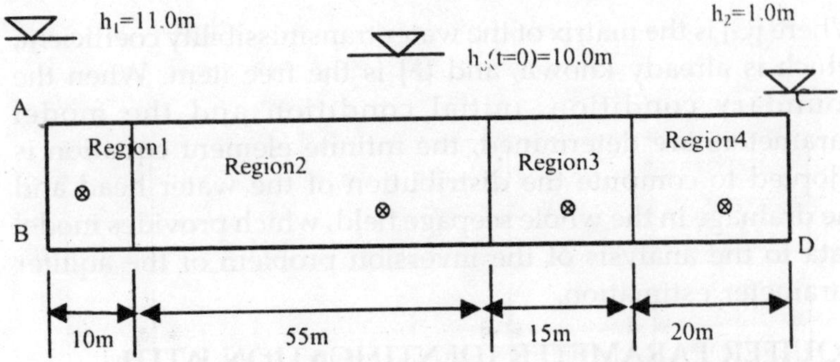

Figure 1: Configuration of the Two-dimensional Groundwater Flow Model

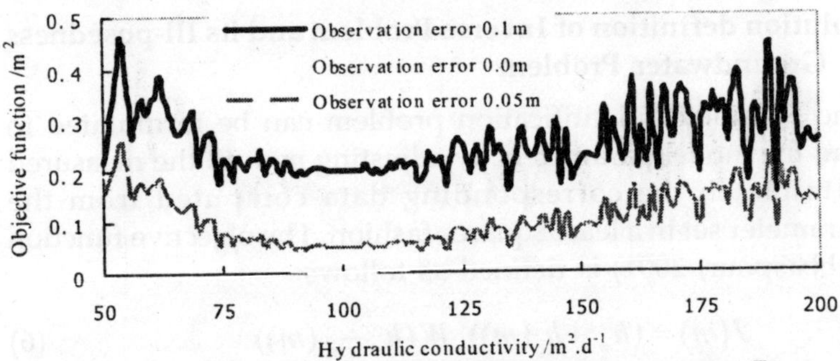

Figure 2: Objective Functions of the Different Measurement Errors

The solution of the inverse problem consists in obtaining a minimum of an objective function which is defined taking into account the mathematical structure of the material model and asset of experimental data. This generally results (Simoni, 2001) in a non-linear programming constrained problem of the form:

$$\min\{J(m,h_m),\quad m\in\Re^P, h_m\in\Re^M; g^j<0 \tag{7}$$

Where *m* represents the variable vector, which belongs to the space of admissible parameters R^P, h_m the vector of measured data, which belongs to the space R^M. g^j are inequality constraints, which define the feasible domain S:

$$S=\{m\in\Re^P, g^j<0\} \tag{8}$$

The constraints can represent physical links between the primary physical variables and the model parameters, information concerning the values of parameters and conditions to guarantee that all mathematical functions involved can be defined and calculated. In the optimization process, the difference between the experimental result and the theoretical prediction is measured by a norm value, here referred to as the individual norm. The individual norms of the tests form an objective function *J(m)* which then gives a scalar measure of the error between the experimental observations and the model predictions. From mathematical point of view (Mattsson, 2001), the optimization problem involves the minimization of the objective function:

$$J(m) \rightarrow \min \tag{9}$$

Where *m* is a vector containing the optimization variables(here model parameters) with the bound constraints:

$$m_l < m < m_u \tag{10}$$

Where m_l and m_u are, respectively, the lower and upper bounds of *m*

Simulated Annealing Algorithm and its Application to Aquifer Parameter Estimation

Simulated annealing (Alkhanmis, 1999) is another important algorithm which is powerful in optimization and high-order problems. It uses random processes to help guide the form of its search for minimal energy states. Simulated annealing is a generalization of a Monte Carlo method for examining the equations of state and frozen states of n-body systems. The concept is based on the manner in which liquids freeze or metals recrystalize in the process of annealing. In an annealing process a melt, initially at high temperature and disordered, is slowly cooled so that the system at any time is approximately in thermodynamic equilibrium. As cooling proceeds, the system becomes more ordered and approaches a "frozen" ground state at *T*=0. Hence the process can be thought of as an adiabatic approach to the lowest energy state. If the initial temperature of the system is too low or cooling is done insufficiently slowly the system may become quenched forming defects or freezing out in meta-stable states, that is, trapped in a local minimum energy state. Simulat*ed annealing* is a very general optimization method which stochastically simulates the slow cooling of a physical system. The idea is that there is a objective function *J*, which associates a objective function with a state of the system, a temperature *T*, and various ways to change the state of the system. The algorithm works by iteratively proposing change and either accepting or rejecting each change. Having proposed a change we may evaluate the change *dJ* in *J*. The proposed change may be accepted or rejected by the *Metropolis* criterion; if the objective function decreases ($dJ<0$), the change is accepted unconditionally; otherwise it is accepted but only with probability $exp(-dJ/T)$. For given old solution, a new solution (Jeong, 1996) can be created as following:

$$m_{new} = m_{old} + \Delta m \tag{11}$$

Where *Dm* is a random perturbation of solution. The accepted probability of the new solution, $p_{new,}$ will be expressed:

$$p_{new} = \begin{cases} 1 & J_{old} \geq J_{new} \\ \exp[-\delta J / T_k] & J_{old} < J_{new} \end{cases} \tag{12}$$

Where $dJ=J_{new}-J_{old}$. Three parameters essential for implementation of the simulated annealing algorithm are as follows:1)initial value of the control parameter, $T_{i,}$ 2)the number of perturbations generated at each *T*, and 3)the decrement of the control parameter *T*. These parameters affect the speed of the algorithm and the quality of the final solution. A simple approach (Alkhanmis, 1999) is to choose a value for *Ti* that allows a large percentage of non-improving solutions to be accepted. The number of solutions generated at each *T* is selected to allow equilibrium to take place before decreasing *T*. The decrement of *T* is chosen such that it allows only small changes in the value of *T*. The equation used for decreasing *T* is expressed as follows :

$$T_{k+1} = \xi T_k \tag{13}$$

Where x=0.9 is a typical selection. A crucial requirement for the proposed changes is *reachability* or *ergodicity*-that there be a sufficient variety of possible changes that one can always find a sequence of changes so that any system state may be reached from any other. When the temperature is zero, changes are accepted only if J decreases, an algorithm also known as hill-climbing, or more generally, the greedy algorithm. The initial temperature (Chen, 2002) can be determined as

$$T_i = \frac{-1}{\ln p_i} \tag{14}$$

Where p_i is the desired initial acceptable probability. It is usually between 0.7and 0.9. similarly, the final temperature can be determined as

$$T_f = \frac{-1}{\ln p_f} \tag{15}$$

Where p_f is the desired final acceptable probability. It is usually very close to zero. The system soon reaches a state in which none of the proposed changes can decrease the objective function, but this is usually a poor optimum. In real life, we might be trying to achieve the highest point of a mountain range by simply walking upwards; we soon arrive at the peak of a small foothill and can go no further. On the contrary, if the temperature is very large, all changes are accepted, and we simply move at random ignoring the cost function. Because of the reachability property of the set of changes, we explore all states of the system, including the global optimum. The system evolves until a stop criterion is reached. Very fast simulated annealing scheme (Rosen, 1995) proposed by Ingber is applied to produce new solution

$$\Delta m_{\mathrm{i}} = \eta_{\mathrm{i}}(m_{\mathrm{imax}} - m_{\mathrm{imin}}) \tag{16}$$

$$\eta_i = sign(u_{\mathrm{i}} - 0.5)T_{\mathrm{i}}[(1 + \frac{1}{T_{\mathrm{i}}})^{|2u_i - 1|} - 1] \tag{17}$$

Where m_{max} and m_{min} represent up and down bounds for parameters, respectively; u_i is a random value in [-1,1) domain, the value of h_i is just located in [-1,1). The whole process for parameter identification using simulated annealing is shown as follows:

Step 1 : Initial parameters(initial T and temperature descent rate a) are fixed.

Step 2 : Initial solution is generated and the corresponding J is calculated.

Step 3 : The system solution is updated according to the mechanism designed.

Step 4 : Parameter T is modified according to the descent rate established.

Step 5 : One comes back to the step 3 to calculate the next solution from the current one up to *T* or *e* reaches a value fixed beforehand.

Step 6 : The best solution visited is written as last solution of inverse problem.

To test the applicability and efficiency of the parameter identification procedure, a two dimensional flow problem is considered, as shown as Fig.1. The aquifer is bound by two constant-head boundaries with the initial heads both at 11m. At t>0, the head at right boundary is instantaneously lowered

Table 1
Hydraulic Conductivity and Specific Storage Coefficient

Parameters	$k_1 / m^2.d^{-1}$	$k_2 / m^2.d^{-1}$	$k_3 / m^2.d^{-1}$	$k_4 / m^2.d^{-1}$	S_s
Values	2000.0	100.0	10.0	1000.0	0.02

Table 2
Measured Water Head Data at Different Points at Different Times

Observation time/d	*Observation point 1# /m*	*Observation point 2#/m*	*Observation point 3#/m*	*Observation point 4#/m*
0.1	10.68716	9.979836	9.170668	3.640028
0.2	10.72888	9.807406	8.396107	3.374715
0.3	10.69714	9.537611	7.756343	3.121658
0.4	10.63557	9.198385	7.028817	1.584894
0.5	10.56086	8.835012	6.398876	1.423336

to 10m. The transient head distribution is simulated using finite element method with a time step of 0.1 day. Hydraulic conductivity and specific storage coefficient are listed in Table 1. Table 2 records measured water head data at different points at different times.

According to the flow mathematical model with finite element method and measured water-head data, the aquifer parameters are identified with simulated annealing algorithm. Fig.3 shows the influence of initial temperature of simulated annealing algorithm to convergence process.

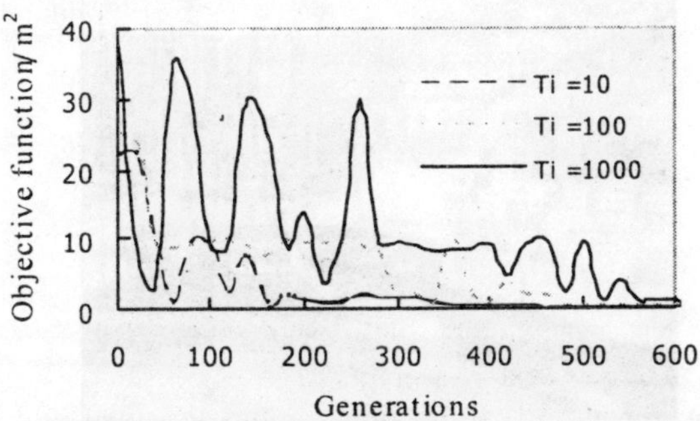

Figure 3: Influence of Initial Temperatures to Convergence Process

Table 3
Comparison of Identified Hydraulic Conductivity and Specific Storage Coefficient with Exact Values

Model parameters	k_1/ $m^2\times d^{-1}$	k_2/ $m^2\times d^{-1}$	k_3/ $m^2\times d^{-1}$	k_4/ $m^2\times d^{-1}$	S_s
Exact Values	2000.0	100.0	10.0	1000.0	0.02
Identified values	2010.3	99.7	10.2	1009.0	0.018

In order to simulate observation errors, the measured water heads can be simulated by adding a random error to the theoretical values

$$h_m^* = h_m + sign(R-0.5)\times\Delta h \tag{18}$$

Where h_m^* are the measured data with observation errors, h_m are the measured data without observation errors. *Sign* is the sign function, and *R* is a random variable in the interval

[0,1], Δh is observation error. Comparison of identified hydraulic conductivity and specific storage coefficient with exact values is listed in Table 3.

PRACTICAL APPLICATIONS OF INVERSION ALGORITHM

Baishan Hydropower Station, as shown in Fig.4, is located in the Second Songhuajiang River in Jilin province, China. It consists of a 149.5-meter-high concrete heavy-pressure dam,

Figure 4: Baishan Hydropower Station

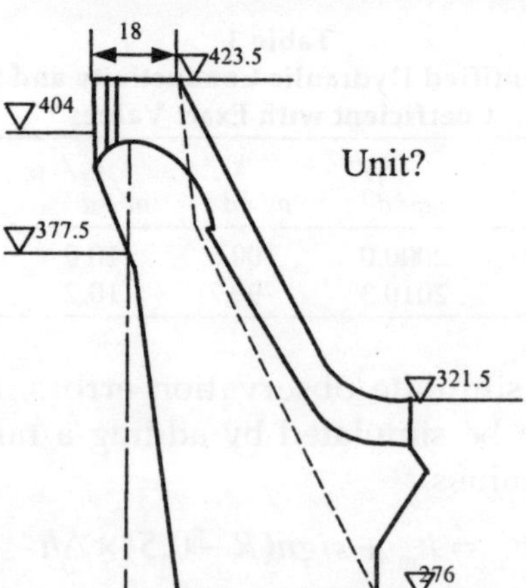

Figure 5: Cross-section of Baishan Dam at Block 18

a weir with four 12´13 meter tunnels on top of the 404-meter-high spillway dam, three 6 × 7 meter tunnels for discharging water are 350 meters high, an underground powerhouse with an installed generating capacity of 900,000 KW and another powerhouse on the surface with an installed generating capacity of 600,000 KW. The dam is 423.5 meters high and the reservoir has a storage capacity of 6.812 billion cubic meters. Its highest normal storage water level is 413 meters. The capacity for water control storage is 3.54 billion cubic meters while the flood control storage capacity is 950 million cubic meters. Cross-section of Baishan dam at block 18 is shown in Figure.5, Figure.6 shows the disposition of observation holes for dam uplift pressure at Block 18.

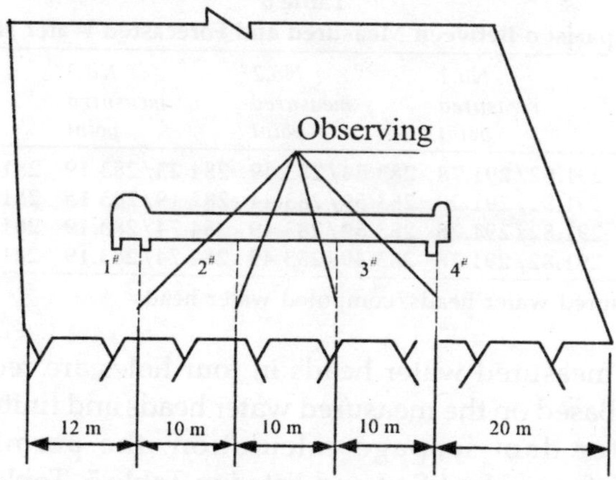

Figure 6: Disposition of Observation Holes for Dam Uplift Pressure at Block 18

In order to identify the permeability coefficients of rock foundation, the three-dimensional finite element model for seepage calculation is carried out. The seepage fields of the dam and its rock foundation at different load cases are computed. According to the prior information of pumping water test in field, the domains of identification parameters are determined. The training sample pairs are got basing on finite element analysis. The rock foundation is divided into 3 sub-regions, rock base before concrete certain, concrete certain and rock base after concrete certain.

Table 4
Measured Data of Water Heads in the Observation Holes

Measuring date	*Upstream water elevation*	*Downstream water elevation*	*water head h_1*	*water head h_2*	*water head h_3*	*water head h_4*
19980910	413.00	290.80	291.82	283.59	284.74	281.72

Table 5
Identification Results of Permeability Coefficients

Rock foundation (I) $k_1/10^{-9}m\times s^{-1}$	*Concrete certain $k_2/10^{-9}m\times s^{-1}$*	*Rock foundation (II)$k_3/10^{-9}m\times s^{-1}$*
44.05	6.16	52.80

Table 6
Comparison Between Measured and Forecasted Water Heads

Measuring date	*No.1 measured point*	*No.2 measured point*	*No.3 measured point*	*No.4 measured point*
19971015	291.82/291.78	283.54/283.49	284.23/283.19	281.72/281.13
19980108	291.31/291.29	283.59/283.44	282.19/283.15	281.82/282.14
19980901	291.82/291.78	283.59/283.49	284.74/283.19	281.72/282.12
19981012	291.82/291.78	283.49/283.49	284.74/283.19	281.72/282.12

Note: measured water heads/computed water heads

The measured water heads in four holes are recorded in Table 4. Based on the measured water heads and finite element model for dam seepage calculation, the permissibility coefficients are identified and listed in Table 5. Table 6 shows the comparison between measured and forecasted water heads with finite element method according to estimated permissibility coefficients.

CONCLUSIONS

Efficient planning and management strategies are essential for optimum utilization of resources and it is considered as a process for continuous improvement and sustainable development. Also it is required to formulate mathematical models and introduce new techniques to solve the same for planning efficient strategies. It is well known that the identification of parameters through optimization is an ill-posed inverse problem, in the sense that it may have multiple optimal solutions. Also small errors in the measured data may propagate and produce misleading solutions. Thus, the optimization method has to incorporate a regularization strategy to control the error propagation, as well. Modern heuristic search techniques, such as genetic algorithm, simulated annealing and ant colony system, are well suited for solving the parameter identification problem in groundwater flow model. The gradient based methods are not applicable for this type of inverse problem because of the difficulty in evaluating the function derivatives and the presence of many local minimum points in the objective function. The mathematical model of underground water flow is introduced as basis to identify the permeability coefficients of rock foundation by observing the water heads of the underground water flow. Based on the finite element model of groundwater flow and the measured water heads of uplift pressure at dam tunnel, simulated annealing algorithm is applied to estimate the permeability coefficients of rock foundations of concrete dam. According to the identified permeability coefficients of rock foundations of concrete dam and finite element analysis, the forecasted water heads can precisely agree with the observed water heads.

ACKNOWLEDGEMENT

This research is funded by the National Natural Science Foundation of China (No. 90815023) and by The National Basic Research Program (973 Program) (No. 2007CB714006).

REFERENCES

Abbaspour, K. C., (2001), 'Estimating Unsaturated Soil Hydraulic Parameters Using Ant Colony Optimization', *Advances in Water Resources*, Vol.24, No.8, pp.827-841.

Alkhanmis, T. M., (1999), 'Simulated Annealing for Discrete Optimization with Estimation', *European Journal of Operational Research*, Vol.116, No.3, pp.530-544.

Ayvaz, M. T., (2007), 'Simultaneous Determination of Aquifer Parameters and Zone Structures with Fuzzy C-means

Clustering and Meta-heuristic Harmony Search Algorithm', *Advances in Water Resources,* vol.30, no.11, pp.2326-2338.

Bau, D. A & Mayer, A. S., (2008), 'Optimal Design of Pump-and-treat Systems Under Uncertain Hydraulic Conductivity and Plume Distribution', *Journal of Contaminant Hydrology,* Vol.100, No.1-2, pp.30-46.

Carrera, J., (1986), 'Estimation of Aquifer Parameters Under Transient and Steady State Conditions', *Water Resources Research,* Vol.22, No.2, pp.199-210.

Chen, T. Y., (2002), 'Efficiency Improvement of Simulated Annealing in Optimal Structural Designs', *Advances in Engineering Software,* Vol.33, No.7-10, pp.675-680.

Eldho, T. I & Rao, B. V., (1997), 'Simulation of Two-dimensional Contaminant Transport with Dual Reciprocity Boundary Elements', *Engineering Analysis with boundary elements,* Vol.20, No.3, pp.213-228.

Jeong, I. K., (1996), 'Adaptive Simulated Annealing Genetic Algorithm for System Identification', *Engineering Applications of Artificial Intelligence,* Vol.9, No.5, pp.523-532.

Kool, J. B. (1987), 'Parameter Estimation for Unsaturated Flow and Transport Models – A Review', *J. Hydrol,* Vol. 91, No.3-4, pp.255-293.

Mattsson, H., (2001), 'Optimization Routine for Identification of Model Parameters in Soil Plasticity', *Int. J. for Numerical and Analysis Methods in Geomechanics,* Vol.25, No.5, pp. 435-472.

Qin, X, S, Huang, G, H & He, L., (2009), 'Simulation and Optimization Technologies for Petroleum Waste Management and Remediation Process Control', Journal of Environmental Management, Vol.90, No.1, pp. 54-76.

Reshmidevi, T. V. Jana, R & Eldho, T, I., (2008), 'Geospatial Estimation of Soil Moisture in Rain-fed Paddy Fields Using SCS-CN-based Model', *Agricultural water management,* Vol.95, No.4, pp.447-457.

Rivest, M, Marcotte, D & Pasquier, P., (2008), 'Hydraulic Head Field Estimation Using Kriging with an External Drift: A Way to Consider Conceptual Model Information', *Journal of Hydrology,* Vol.361, No.3-4, pp.349-361.

Rosen, I. I. (1995), 'Comparison of simulated annealing algorithms for conformal therapy treatment planning', *Int. J. Radiation Oncology Biol. Phys*, vol.33, no.5, pp.1091-1099.

Simoni, L. (2001), 'An Accelerated Algorithm for Parameter Identification in a Hierarchical Plasticity Model Accounting for Material Constraints', *Int. J. for Numerical and Analysis Methods in Geomechanics*, Vol.25, No.3, pp.263-272.

Tan, C. C, Tung, C, P & Chen, C, H, (2008), 'An Integrated Optimization Algorithm for Parameter Structure Identification in Groundwater Modeling', *Advances in Water Resources*, Vol.31, No.3, pp.545–560.

Wang, P. P, (1998), 'An Efficient Approach for Successively Perturbed Groundwater Models', *Advances in Water Resources*, Vol.21, No.6, pp.499-508.

Yeh, W, (1986), 'Review of Parameter Identification Procedures in Groundwater Hydrology: The Inverse Problem', *Water Resources Research*, Vol.22, No.2, pp.95-108.

Young, D. L. Eldho, T, I & Chang J, T. (2006), 'Large Eddy Simulation of Turbulent Flows in External Flow Field Using Three-step FEM–BEM Model', *Engineering Analysis with Boundary Elements*, Vol.30, No.7, pp. 564-576.

Zadeh, K. S. (2008), 'Parameter Estimation in Flow through Partially Saturated Porous Materials', *Journal of Computational Physics*, Vol.227, No.24, pp.10243-10262.

Zheng, C. (1996), 'Parameter Structure Identification Using Tabu Search and Simulated Annealing', *Advances in Water Resources*, Vol.19, No.4, pp.215-224.

12

BIOSORPTION OF CHROMIUM (VI) BY *TAMARINDUS INDICA* POD SHELLS

N. Ahalya, R. D. Kanamadi & T. V. Ramachandra

INTRODUCTION

Rapid industrialisation has led to increased disposal of heavy metals into the environment and hence effluent treatment is one of the most important targets for industry. Different alternatives for treating effluents are described in literature, including chemical precipitation, carbon adsorption, ion exchange and membrane separation process, among others (Juang and Shiau, 2000; Lacour *et al.*, 2001; Yan and Viraraghavan, 2001). The most common of these alternatives is chemical precipitation. However some limitations in the process can be pointed out, such as high cost, low efficiency, labour intensive operation, and lack of selectivity of the precipitation process (Lee *et al.*, 1998). On the other hand, the use of other natural materials such as peanut shells, soybean hulls and corncobs, which are available in large quantities, may present high potential as inexpensive sorbents for effluent treatment (Marshall *et al.*, 1999; Wafwoyo *et al.*, 1999; Vaughan *et al.*, 2001). Bailey *et al.*, (1999) have reviewed the potential of a wide variety of low cost sorbents for heavy metals. A low cost sorbent is the one that requires less processing, is abundant in nature or is either a byproduct or waste material from another industry. These materials could be an alternative for expensive treatment processes.

In India, tamarind (*Tamarindus indica* L.) is an economically important tree which grows abundantly in the dry tracts of Central and South Indian States. Indian production of tamarind is about 3 lakh (0.3 million) tonnes per year. The hard pod shell is removed (deshelled) when the fruit is ripe and the fruit is the chief acidulant used in the preparation of foods. The shells are discarded as waste and since it is available free of cost, only the transport cost is involved for hauling it from the point of generation for wastewater treatment. Hence, recycling of this waste for wastewater treatment would not only be economical but also help to solve waste disposal problems.

Hexavalent chromium frequently encountered in electroplating effluent, is one of the most toxic heavy metal in the environment and is also a known carcinogen and mutagen in humans and animals (Zhao and Duncan 1998; Salunkhe *et al.*, 1998). The possibility of using Tamarind shells, an agricultural waste for the removal of Cr (VI) from aqueous solutions has been examined in this paper.

MATERIALS AND METHODS

Tamarind pod shells were obtained from a de-hulling unit in Bangalore, Karnataka, India. The tamarind shells were washed thoroughly with distilled water and boiled to remove colour and impurities. They were dried in a hot air oven at 105°C and ground to a uniform size of 0.8 mm. So prepared tamarind pod shell particles were used for further studies. The stock solution (1000 mgl^{-1}) of Cr (VI) was prepared using potassium dichromate salt. Biosorption studies were performed in batch process. The processed tamarind shells equivalent to 1g dry weight were added to 100 ml of an aqueous solution of chromium (pH 2.0; adjusted with 0.1 M H_2SO_4) of required concentration. The concentration of the unadsorbed chromium was determined spectrophotometrically at 540 nm using the diphenylcarbazide reagent. The effect of several parameters such as pH, metal concentration, contact time and quantity of biosorbent on adsorption is studied. The pH of the chromium aqueous solution was adjusted using 0.1 N HCl/NaOH. To study the functional groups responsible for the biosorption of Cr (VI), the biomass (tamarind pod shells) was analyzed using a Fourier transform infrared spectrometer (FTIR). Adsorption of Cr (VI) on the walls of the glassware is found to be negligible and is determined by running blank experiments.

RESULTS AND DISCUSSION

Effect of Agitation Time and Initial Cr (VI) Concentration

The equilibrium time required for biosorption of Cr (VI) on tamarind shells is obtained by studying adsorption of Cr (VI) at various initial concentrations. The data is represented graphically in Figure 1. The extent of adsorption efficiency increases sharply with time and attains equilibrium at about 60 minutes for all the concentration studied. According to the above results, the agitation/equilibrium time was fixed at 120 minutes for the rest of the experiments so that complete equilibrium is reached. The gradual increase in the rate of biosorption of Cr (VI) and plateau thereon after 60 minutes indicates that the adsorption occurs through a smooth continuous formation of adsorption layer till saturation.

Effect of pH

The biosorption of Cr (VI) was found to depend on the pH of the metal solution (Figure 2). Biosorption capacity was found to decrease with an increase in the pH, maximum adsorption being observed at a pH of 2. It is well known that the dominant form of Cr (VI) at this pH is $HCrO_4^-$. Increasing the pH will shift the concentration of $HCrO_4^-$ to other forms, CrO_4^{2-} and $Cr_2O_7^{2-}$. It can be concluded that the active form of Cr (VI) that can be adsorbed by tamarind pod shell is $HCrO_4^-$.

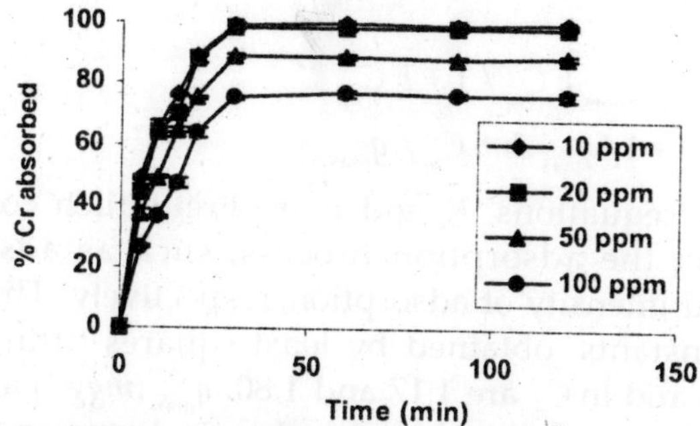

Figure 1: Percentage Biosorption of Chromium (VI) from Solution of Different Concentration, pH 2.0, by 10 gl^{-1} Biomass as Related to the Time of Contact at 120 rpm

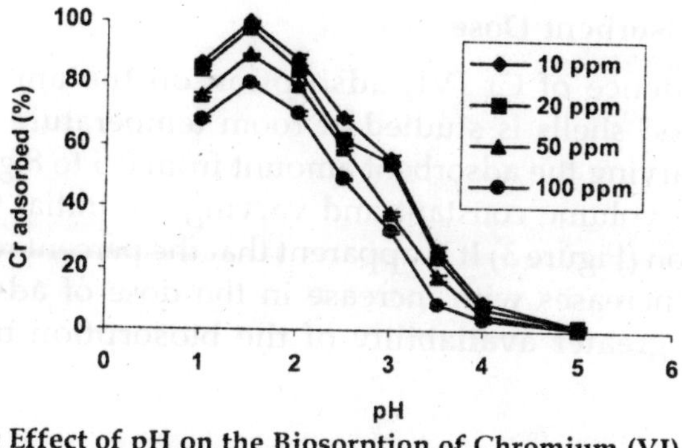

Figure 2: Effect of pH on the Biosorption of Chromium (VI) at Different Concentrations, by 10 gl^{-1} at 120 rpm with Equilibrium Time of 120 min

Adsorption Isotherms

Among various plots employed for analyzing the nature of adsorbate-adsorbent interaction, adsorption isotherm is the most significant. The results of adsorption studies of chromium (VI) at different concentrations ranging from 20 to 500 ppm on a fixed amount of adsorbent are expressed by two of the most popular isotherm theories viz., Freundlich and Langmuir isotherms. These isotherm equations are as follows:

Freundlich:

$$q = K_f C_{eq}^{1/n}$$

$$\ln q = \ln K_f + 1/n \ln C_{eq}$$

Langmuir

$$q = q_{max}\, b\, C_{eq} / 1 + b\, C_{eq}$$

$$C_{eq}/q = 1/q_{max}.b + C_{eq}/q_{max.}$$

In the above equations, K_f and n are Freundlich constants, which affect the adsorption process, such as adsorption capacity and intensity of adsorption, respectively. The values of these constants, obtained by least squares fitting of the data on $\ln q$ and $\ln C_{eq}$ are 1.17 and 1.80. q_{max} (mgg^{-1}) and b are Langmuir constants related to monolayer adsorption capacity and energy of adsorption respectively. The values of the parameters, evaluated by the least square fitting of the data on C_{eq} versus C_{eq}/q are 27.73 and 0.008.

Effect of Adsorbent Dose

The dependence of Cr (VI) adsorption on the amount of tamarind pod shells is studied at room temperature and at pH2.0 by varying the adsorbent amount from 0.5 to 8 g, while keeping the volume constant and varying the initial Cr (VI) concentration (Figure 3) It is apparent that the percent removal of Cr (VI) increases with increase in the dose of adsorbent due to the greater availability of the biosorption binding sites.

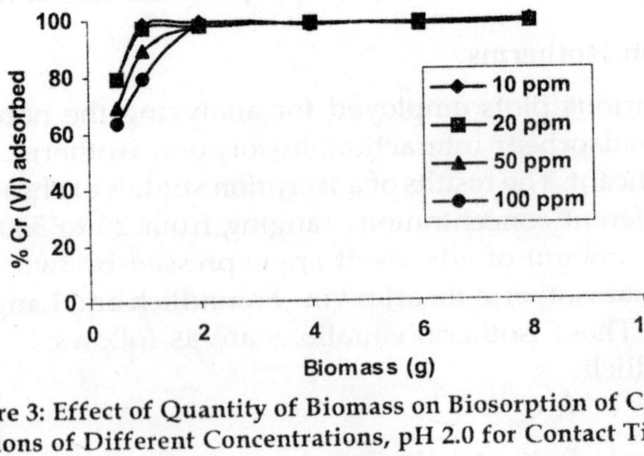

Figure 3: Effect of Quantity of Biomass on Biosorption of Cr (VI) from Solutions of Different Concentrations, pH 2.0 for Contact Time 120 min at 120 rpm

Infrared Spectroscopy

An un-reacted tamarind pod shell sample and pretreated with 100 mg/l Cr (VI) solution were analysed using FTIR, to determine the functional groups that play an important role in the biosorption of Cr (VI). The wavenumbers along with their corresponding functional groups are given in Table 1. The infrared spectroscopic studies show that several functional groups are available on the surface of the adsorbent for binding Cr (VI) ions.

Table 1
IR Absorption Bands and the Corresponding Functional Groups

Wavelength (cm^{-1})	*Functional group*
3425.30	– OH, – NH
2925	– CH
1745.50	– COO –
1621.98	C==C, – OH
1517	C==C
1461.42	C==C in carbon rings or CH from methyl
893.25	– CH

Desorption Studies

The desorption of Cr (VI) from previously Cr (VI) loaded tamarind pod shells using distilled water was attempted. But Cr (VI) desorption was not observed. Hence, experiments were conducted with acid and alkali solutions to desorb Cr (VI) ions. The results obtained (not shown) indicate that the desorption of Cr (VI) ions with acid was not achieved even when 0.1 N HCl and 0.2 N HCl were used. However, there was little desorption with alkali solutions. It was found that Cr (VI) desorption was 27.8% with 0.2 N NaOH. The results of desorption studies indicate that either chemisorption or ion-exchange as the possible mechanism for Cr (VI) binding on the tamarind pod shells.

CONCLUSION

This is the first time in literature that tamarind pod shells have been used for the adsorption of Cr (VI). As the tamarind pod shells are easily available, its utility as biosorbent will be economical and be viewed as part of waste management strategy. Containing very less amount of protein (Nitrogen = 0.94%), tamarind pod shells is advantageous over the protein rich algal and fungal biomass projected as metal biosorbents, since proteinous materials are likely to putrefy under moist conditions.

REFERENCES

Bailey S. E., Olin T. J., Bricka R. M., and Adrian D. D. (1999), A Review of Potentially Low Cost Sorbents for Heavy Metals, *Water Res.* 33: 2469-2479.

Juang R. S., and Shiau R. C. (2000), Metal Removal from Aqueous Solution Using Chitosan Enhanced Membrane Filtration, *J. Membrane Sci.*, 165: 159-160.

Lacour S., Bollinger J. C., Serpaud B., Chantron P. and Arcos R. (2001), Removal of Heavy Metals in Industrial Wastewaters by Ion Exchanger Grafted Textiles, *Anal. Chim. Acta*, 428: 121-132.

Lee S. H., Jung C. H., Chung H., Lee M. Y., and Yang J. (1998), Removal of Heavy Metals from Aqueous Solutions by Apple Residues, *Process Biochem.*, 33: 205-211.

Marshall W. E., Wartelle L. H., Boler D. E., Johns M. M., and Toles C. A. (1999), Enhanced Metal Adsorption by Soybean Hulls Modified with Citric Acid, *Bioresour. Technol.*, 69: 263-268.

Salunkhe P. B., Dhakephalkar P. K., and Paknikar K. M. (1998), Bioremediation of Hexavalent Chromium in Soil Microcosms, *Biotechnol. Lett.*, 20(8): 749-751.

Vaughan T., Seo C. W., and Marshall W. E. (2001), Removal of Selected Metal Ions from Aqueous Solution Using Modified Corncobs, *Bioresour. Technol.*, 78: 133-139.

Wafwoyo W., Seo C. W., Marshall W. E. (1991), Utilisation of Peanut Shells as Adsorbents for Selected Metals, *J. Chem. Technol. Biotechnol.*, 74: 1117-1121.

Yan G., and Viraraghavan T. (2001), Heavy Metal Removal in a Biosorption Column by Immobilised *M. Rouxii* Biomass, *Bioresour. Technol.*, 78: 243-249.

Zhao M., and Duncan J. R. (1998), Column Sorption of Cr (VI) from Electroplating Effluent Using Formaldehyde Cross-Linked Sacchromyces Cerevisae, *Biotechnol. Lett.* 20(6): 603-606.

13

AQUACULTURE IMPACTS ON GROUND WATER QUALITY IN PENNER RIVER BASIN USNIG REMOTESENSING & GIS: A MODEL STUDY

S. S. Asadi, K. Kiran, G. Rajani & M. Anji Reddy

INTRODUCTION

Mankind depends upon natural resources to survive, this unavoidable depency has resulted in exploitation of natural resources over the centuries. And as the population of the world is increasing rapidly the rate at which the natural resources are depleting is increasing dangerously. This problem calls for an effective management of community depends upon how effectively it manages and uses its natural resources.

Demographic explosion accelerated urbanization, aquaculture, salt cultivation and excessive withdrawl of ground water resulted in huge stress on ground water in coastal area. Hence, delineating fresh water zones from intercalated saline zones of coastal tracts needs. Pennar in Nellore district is no exemption of this type of problem.

Andhra Pradesh being predominantly an agricultural state. Necessitates the practice of conjunctive use of surface and ground water in its deltas. Pennar in Nellore district is no exemption of this type of problrm.Salt water intrusion which refers to the movement of salty water into fresh water aquifers or to the encroachment of saline water into fresh water estuaries has been a major ground water resource problem in coastal environments for decades. Increasing population in Nellore and its surrounding areas rapidly increased aquaculture, salt cultivation are resulting in ground water consumption which leads to the problem as the salt water moves further inland in response to pumping. This critical problem requires proper management to protect the ground water resources.

DESCRIPTION OF THE STUDY AREA

The study area extends from the Nellore ayacut on Pennar upto Bay of Bengal about 50 kms downstream of Nellore.The area is included in SOI Toposheets 57N14, 15 and 66B2, B3. The study area is underlain by alluvium of recent age. It includes all the areas irrigated by Pennar River water and ground water towards Northern bank of Pennar River. The ground water in the area is abstracted through filter points mostly and also through the bore wells and dug wells. The over extraction of the ground water may lead to salt water intrusion

Objectives

- To establish monitoring network for determining the quality of surface and shallow ground water.
- To identify potential coastal aquifers and mapping of these aquifers in order to assess the ground water resource and ground water draft.
- To monitor chemical quality and prepare depth-wise-quality map to show salinity distribution at various levels.
- Specific aim of the study is to assess the impact of existing proposed human activities on saltwater intrusion and to

develop seawater-freshwater interface model to predict the intrusion.

METHODOLOGY

Data Collection

Different data products required for the study include Survey of India (SOI) toposheets bearing with numbers 56O/14 and 56O/15 on 1 : 50,000 scale. Fused data of IRS-1D PAN and LISS-III satellite imagery obtained from National Remote Sensing Agency (NRSA), Hyderabad, India. Collateral data collected from related organizations, comprises of water quality and demographic data.

Database Creation

Satellite imageries are georeferenced using the ground control points with SOI toposheets as a reference and further merged to obtain a fused, high resolution (5.8 m of PAN) and colored (R, G, B bands of LISS-III) output in EASI/PACE Image processing software. The study area is then delineated and subsetted from the fused data based on the latitude and longitude values and a final hard copy output is prepared for the generation of thematic maps using visual interpretation technique as shown in Figure 1. These thematic maps (raster data) are converted to vector format by scanning using an A0 flatbed deskjet scanner and digitized using AutoCAD software for generation of digital thematic maps using Arc/ Info and ARCVIEW GIS software. The GIS digital database consists of thematic maps like land use/land cover, drainage, road network using Survey of India (SOI) toposheets and fused data of IRS-ID PAN and IRS-ID LISS-III satellite imagery.

Spatial Database

Thematic maps like base map and drainage network maps are prepared from the SOI toposheets on 1 : 50,000 scale using AutoCAD and Arc/Info GIS software to obtain a baseline data. Thematic maps of the study area was prepared using visual interpretation technique from the fused satellite imagery (IRS-ID PAN + IRS-ID LISS-III) and SOI toposheets along with ground truth analysis. All the maps are scanned and digitized to generate a digital output (Figure 1).

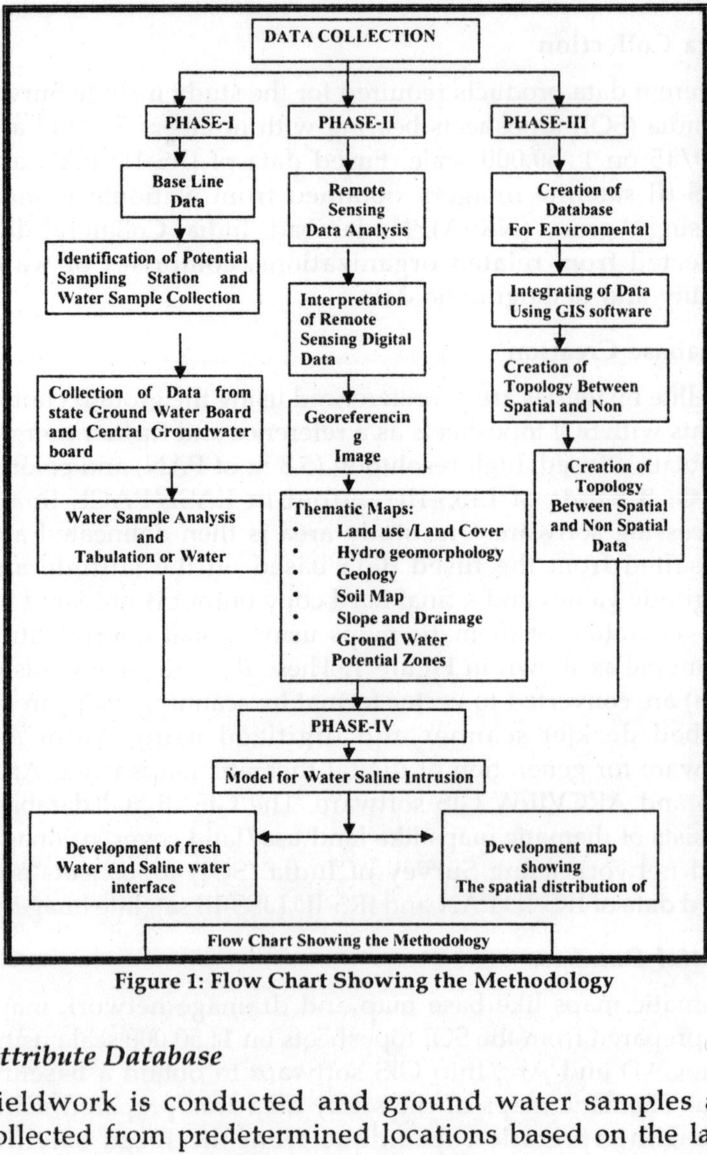

Figure 1: Flow Chart Showing the Methodology

Attribute Database

Fieldwork is conducted and ground water samples are collected from predetermined locations based on the land use and drainage network maps in the study area. Care is taken in collecting the water samples for uniform distribution and density of sampling locations. The water samples were analyzed for various physico-chemical parameters adopting standard protocols (APHA, AWWA, WPCF 1998). The water quality data thus obtained forms the attribute database for the present study (Table 1).

Integrated Data

Based on the above spatial data and Attribute data integrated to create the Model for Saltwater Intrusion. The integrated data Development map showing the zone of saltwater intrusion in coastal aquifers. An interface between fresh water and saline water will be developed.

RESULTS AND DISCUSSION

Base Map

A topographic map is a representation of the shape, size, position and relation of the physical features of an area (IMSD Technical Guidelines 1995). The base map is prepared using SOI toposheet on 1 : 50,000 scale and updated with the help of satellite imagery. It consists of various features like the road network, settlements, water bodies, canals, railway track, vegetation etc. delineated from the toposheet. The map thus drawn is scanned and digitized to get a digital output. The information content of this map is used as a baseline data to finalize the physical features of other thematic maps.

Drainage Network Map

Drainage map is prepared by using Survey of India Topographic maps on 1 : 50,000. All the streams and tanks existing in the study area are marked in this map. This Drainage network map is to know the drainage pattern and for calculating various characteristics like drainage density, basin slope etc. The Pennar and its tribulations do not have enough flow to support direct abstraction for larger towns, so infiltration wells are use for most of the riverside town.

The Pennar Delta is having a four major, fourteen medium and forty-one minor drains. The main drain basins are Maldevi, the Alluri swamp, Nakklavagu and Paideru vagu. The Paideru traversing for distance of 43.2 Km joins the Sea at Ponnapudi village at Vidavaluru mandal. The rest three drains, River Pennar, Chippaleru and Uppateru, which finally empties into sea.

Slope Map

Slope classes 1, 2 and 3 are observed in the study area. Most of the study area is covered by nearly level, slope class (94%). Small part of the study area (6%) comes under very gently sloping class of the study area (IMSD Technical Guidelines, 1995). Slope aspect and altitude are important parameters from land utilization point of view. slope is very vital one for land irrigibility and land capability assessment.

Land Use/Land Cover Map

Comprehensive information on land use/land cover is the basic prerequisite for land resources evaluation, assessment, utilization and management. The land use/land cover Categories are Observed such as Standing crop (Paddy), Kherif Un irrigated, Aqua culture, built up land, Brackish water, Salt Ponds, Land with Scrub, land without scrub and others. Paddy is the major agricultural crop in the area. Land use/ Land cover classes encountered in the study area During the Observation of Satellite data the major land use changes are as follow:

- Changes from paddy to shrimp farming,
- Changes from scrub to shrimp farming,
- Changes from paddy to abandoned paddy.

Geomorphology Map

The Geomorphology in the present study area clearly shows the Coastal environment. The main categories are Beach, Coastal plain, Channel islands observes in River mouth and brackish water areas, Salt flats in Brackish water andSides of

Table 1
Water Quality Data

S.No	Sampling Location	Ph	E.C umho/cm	Salinity	Alkalinity mg/l	Sodium mg/l	Potassium mg/l	Total Hardness mg/l	Chlorides mg/l	Calcium mg/l	Magnisiu mg/l	Sodium Absorb-tion Ratio	Tota Dissolved' Solids mg/l
1	Isakapalli	8	2790	3	870	149	101	300	425	90	210	2	1813.5
2	Ragadachilaka	8	649	0	320	44	7	120	35	36	84	1	421.85
3	Khammapalem	8	1115	0	370	125	4	160	241	60	100	2	724.75
4	Allur Road near Rail	8	1667	2	780	354	29	40	199	10	30	15	1083.55
5	Salt pond	7	1012	100	290	13077	1470	54000	91602	5000	49000	16	657.8
6	Salt pond	8	40900	100	340	50769	3676	55000	91602	4200	50800	62	26585
7	Salt pond	7	80500	100	340	13284	1590	34000	91602	5200	28800	20	52325
8	Prawnpond (salt pon	7	18060	11	290	10769	661	2780	91402	2040	740	43	11739
9	Allur N H	8	1354	1	330	122	2	220	404	66	154	2	880.1
10	Allur NH	8	622	0	160	49	5	200	170	60	140	1	404.3
11	Ulvapalli	8	497	0	290	51	1	100	128	45	55	1	323.05
12	Peddapalem	7	737	4	210	76	33	160	177	70	90	1	479.05
13	Singarupalli	8	1265	3	310	125	27	160	312	48	112	2	822.25
14	Peddapalem	7	321	4	220	39	22	80	57	15	65	1	208.65
15	Peddapalem Road	8	293	4	190	42	21	80	57	24	56	1	190.45
16	Allur Town	7	1007	1	550	98	25	200	220	60	140	2	654.55
17	Inupur	8	3490	2	410	593	8	320	964	96	224	8	2268.5
18	Purini	8	6450	4	390	729	39	1120	2021	680	440	5	4192.5
19	Gogulapalli Road	8	7380	5	840	927	44	440	91402	240	200	10	4797
20	Gogulapalli	8	1459	0	590	122	15	340	1914	220	120	1	948.35
21	Sunnapubatti	8	680	0	450	68	25	280	135	84	196	1	442
22	Gugulapalli (East)	8	1172	2	310	102	21	140	305	42	98	2	761.8
23	Gogulupalli	8	5500	3	310	679	15	1000	1680	360	640	5	3575
24	Near Ramatheertham	8	1747	3	450	290	7	200	468	75	125	5	1135.55
25	Mopuru	7	2420	1	310	108	9	580	596	240	340	1	1573

Contd.

S.No	Sampling Location	Ph	E.C umho/cm	Salinity	Alkalinity mg/l	Sodium mg/l	Potassium mg/l	Total Hardness mg/l	Chlorides mg/l	Calcium mg/l	Magnisiu mg/l	Sodium Absorb-tion Ratio	Tota Dissolvedl Solids mg/l
26	Racherlapadu	8	362	0	290	42	8	120	57	50	70	1	235.3
27	Dandigunta	8	3170	1	600	586	7	160	752	48	112	12	2060.5
28	Towards Chakacherla	8	1077	1	500	122	3	160	213	48	112	2	700.05
29	Dandigunta	8	1180	1	400	171	11	180	333	70	110	3	767
30	Gadidladine	8	1688	1	370	387	7	180	518	70	110	7	1097.2
31	Gadidladinne	8	1688	1	450	500	7	200	787	60	140	9	1097.2
32	Chekacherla Cheruvu	7	893	1	340	110	14	160	163	60	100	2	580.45
33	Chandrasekharapura	7	251	0	100	37	1	20	57	6	14	2	163.15
34	Vavalla Road	8	1925	1	800	369	43	120	553	50	70	8	1251.25
35	Padmanabuni Satram	8	1857	1	520	108	2	340	489	102	238	1	1207.05
36	Guvvagunta	8	1083	0	450	145	10	120	262	36	84	3	703.95
37	Pallipalem	8	1420	0	310	136	2	140	248	60	80	3	923
38	Ramco seed company	8	1160	1	550	107	2	220	234	66	154	2	754
39	Kodavaluru	8	2480	1	410	111	91	160	418	140	20	2	1612
40	Gundammapalem	8	1349	1	330	107	3	260	369	200	60	1	876.85
41	Bodduvaripalem	8	1137	0	460	151	4	140	220	42	98	3	739.05
42	Bodduvaripalem	7	1801	2	330	328	144	500	659	150	350	4	1170.65
43	Ramudupalem	8	2669	4	660	702	28	440	1950	240	200	8	1734.85
44	Kudithipalem	8	5960	4	410	563	142	460	2375	260	200	6	3874
45	Kudithipalem Limits	8	765	2	250	72	19	200	45801	75	125	1	497.25
46	Kudipadu	8	3880	1	530	487	8	260	1014	200	60	6	2522
47	Muduvarthipalem	8	885	5	250	65	5	80	128	15	65	2	575.25
48	Pallipalem	8	1466	1	540	137	11	200	326	60	140	2	952.9
49	Muduvarthipalem	7	1306	2	630	127	4	140	269	42	98	3	848.9

Contd.

S.No	Sampling Location	Ph	E.C umho/cm	Salinity	Alkalinity mg/l	Sodium mg/l	Potassium mg/l	Total Hardness mg/l	Chlorides mg/l	Calcium mg/l	Magnisiu mg/l	Sodium Absorb-tion Ratio	Tota Dissolvedl Solids mg/l
50	Pallipalem	7	1384	1	480	143	15	100	546	30	70	4	899.6
51	Nagaraju Thopu	7	1128	0	600	99	7	220	355	66	154	2	733.2
52	Nagaraju Thopu	8	612	0	460	75	9	140	128	50	90	2	397.8
53	Pattapalem	8	7080	5	560	414	27	1180	1560	740	440	3	4602
54	Ganga Patnam	8	1448	0	540	141	9	240	603	72	168	2	941.2
55	Narsapur	8	1555	1	420	112	22	320	347	260	60	1	1010.75
56	Nagaraju Thopu	8	807	0	310	86	11	160	142	65	95	2	524.55
57	Near Marripadu Road	8	1234	1	470	133	5	180	227	70	110	2	802.1
58	Mothalu	8	492	4	430	466	7	440	652	200	240	5	319.8
59	Pallipadu	7	670	0	270	77	9	160	121	48	112	2	435.5
60	Devispet	7	506	0	250	46	3	140	92	55	85	1	328.9
61	High Depth	8	753	0	320	75	21	200	128	60	140	1	489.45
62	Near Podurupadu	8	752	0	400	74	5	240	135	72	168	1	488.8
63	Podugupadu Irrgi.	8	884	0	330	90	7	140	128	42	98	2	574.6
64	Indukurpeta	7	942	0	340	100	2	160	149	48	112	2	612.3
65	Kotturu	8	950	0	350	96	65	220	170	200	20	1	617.5
66	Penna River Bank	8	1012	0	390	77	10	300	206	90	210	1	657.8
67	Indukuru Limits	8	809	0	410	85	5	200	99	80	120	1	525.85
68	Narayanaredd	8	556	0	390	85	7	120	85	50	70	2	361.4
69	Prasanthinagar	8	520	0	260	84	3	140	78	42	98	2	338
70	Navalakura Thota	8	992	0	330	98	5	300	199	90	210	1	644.8
71	Kothakaluva	8	1213	0	480	109	7	200	199	60	140	2	788.45

Contd.

S.No	Sampling Location	Ph	E.C umho/cm	Salinity	Alkalinity mg/l	Sodium mg/l	Potassium mg/l	Total Hardness mg/l	Chlorides mg/l	Calcium mg/l	Magnisiu mg/l	Sodium Absorb-tion Ratio	Tota Dissolved Solids mg/l
72	Kothakaluvu	8	496	0	350	66	8	120	71	50	70	1	322.4
73	Near Narkur Road	8	1583	1	530	199	7	140	312	42	98	4	1028.95
74	Allipuram after locks	8	558	0	330	66	11	180	92	54	126	1	362.7
75	Near Allipuram Road	8	723	0	350	53	8	120	71	50	70	1	469.95
76	Narkur Cross Roads	8	1314	1	500	115	12	440	241	420	20	1	854.1
77	Allipuram	7	475	0	260	68	11	140	85	65	75	1	308.75
78	Nagaraju Thopu	8	765	0	330	79	11	185	128	56	130	2	497.25
79	Ramco Seed Comp.	8	1352	0	492	104	25	184	292	78	106	2	878.8
80	Vavalla Road	8	1186	0	590	116	16	160	312	54	107	3	770.9
81	Kudipadu	8	1485	3	470	279	15	156	454	54	102	5	965.25
82	Mopuru	8	583	1	360	96	9	255	257	100	156	2	378.95
83	Singarupalli	8	1204	2	513	182	27	160	215	47	113	6	782.6
84	Indupur	8	2811	2	437	383	20	413	803	199	214	5	1827.15
85	Allur Town	8	1403	2	530	107	40	197	271	62	135	2	911.95
86	Isakapalli	8	1899	4	710	124	63	250	323	75	175	2	1234.35
87	Muduvarthipalem	8	958	1	465	97	15	180	195	81	99	2	622.7
88	Indukuru Limits	8	748	3	393	217	5	267	300	109	157	3	486.2
89	Ulvapalli	8	1035	0	390	144	9	140	225	45	95	1	672.75
90	Racherlapadu	8	461	1	268	55	6	129	102	48	81	2	299.65
91	Ramco Seed Comp	8	1592	1	473	123	32	173	291	83	91	2	1034.8
92	Ramco Seed Comp	8	1151	0	435	101	15	180	215	65	116	0	748.15
93	Penna River Bank	8	817	0	370	84	8	187	140	61	126	2	531.05
94	Isakapalli	8	2345	4	790	137	82	275	374	83	193	2	1524.25

Contd.

95	Ramatheertham Road	8	2814	2	490	389	8	230	741	138	93	0	1829.1
96	Indukurpeta	8	840	4	400	225	5	300	378	115	176	0	546
97	Devispet	8	794	3	397	221	5	284	339	112	167	3	516.1
98	Devispet	8	876	2	375	124	93	4	180	64	116	2	569.4
99	Peddapalem Road	8	2643	4	870	150	100	286	410	90	210	28	1717.95

River has Alluvium Plain Deep and then Alluvium Plain Moderate is the major geomorphology class of the study area.

Geology Map

The study area geology constitutes mainly Unclasified crystallines, Closepet Granite, Dharwar supergroup, Cuddapah supergroup, Decan traps, Kurnool super group, Rajamahendri sand stone.

Soil Map

The specific objectives of soil mapping are identification, characterization and classification of the soils of the area. The soil types identified in the study area are Coastal sands, Red Loamy soils, Red Earth (Chalkas), Black Cotton soils (light-moderate), Black Cotton soils (Heavy–Deep), Saline soils, Laterite Soils, Red Earth with clayey sub soil. The Pennar delta plain extends from north of Allur to Muthukur in the South over an area of 1470 km^2 with 54 long coast line in the east. The delta has a cuspate shape. At present, the river exhibits no conspicuous delta growth. Due to semi-arid climate prevailing, resulting in the reduction of water discharge and sediment load. Clay and Silt occupied on both sides of the river, Fine sand and clay and silt found in left bank of the river and Coastline occupies sand with shell fragments, silt clay.

Groundwater Potential

The groundwater potential map is prepared based on the analysis of various themes such as geomorphology, land use/ land cover, lineament, intersection points, drainage pattern, lithological evidences by using converging evidence concept, besides the collateral data obtained from State Groundwater Board with necessary field checks. The groundwater potential map reveals the availablequantum of groundwater and is delineated into zones showing high (53%), medium (30%), low (17%), groundwater potential areas (Krishnamurthy, 1996). In the study area High, Medium, Low zone areas water yield ranges > 500 l pm, 300-500 l pm, < 300 l pm has been observed.

Physiography

The Physiography is prepared using the contours derived from Survey of India topo sheets. Their physiography categories

Figure 2

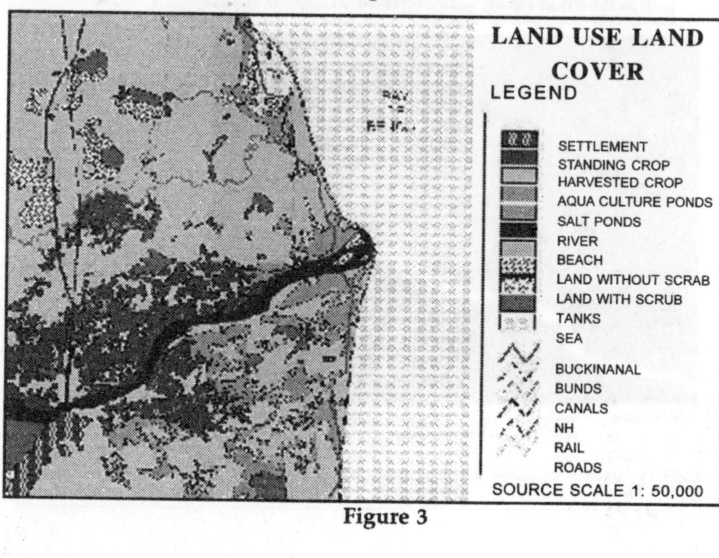

Figure 3

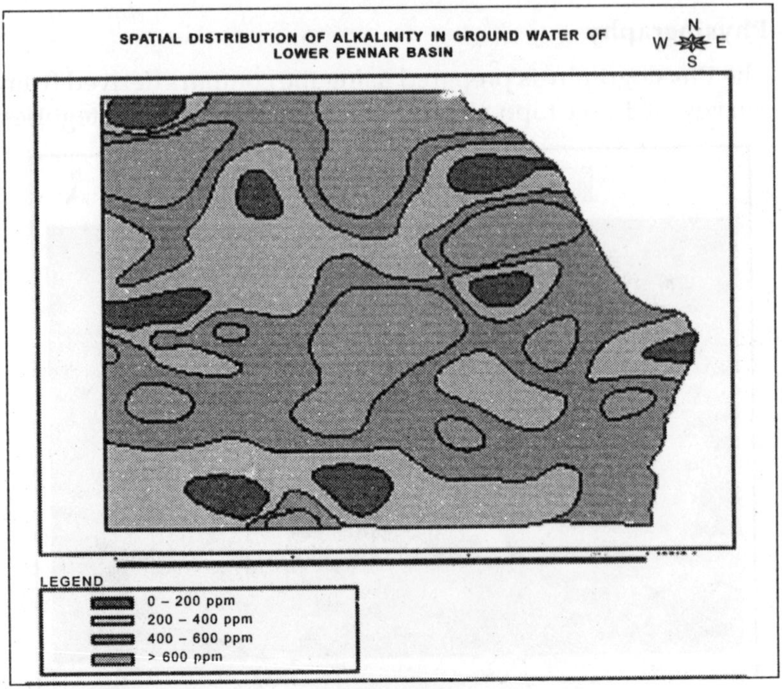

Figure 4

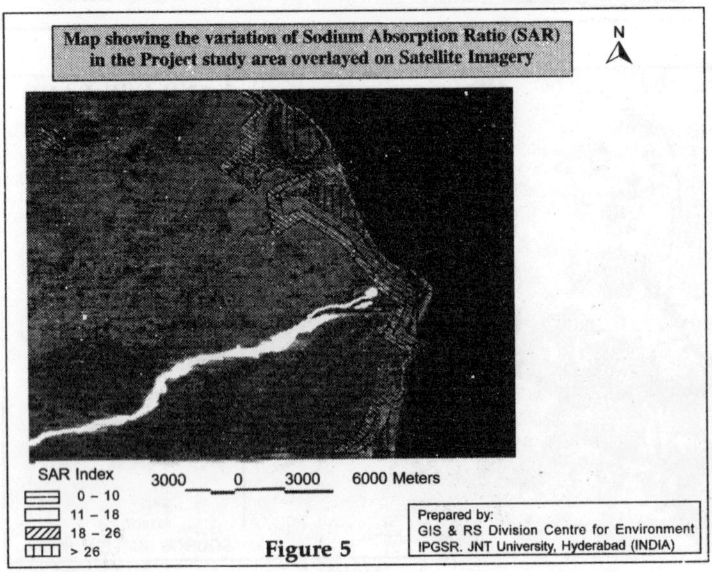

Figure 5

are demarcated in the map as Plains, Undulating land and Hills. The plains occupy the major part of the project area, which indicates nearly level, gently sloping and very gently sloping terrain. The next major unit is the undulating terrain indicating moderately sloping terrain, which indicates the pediments and inselbergs of the geomorphic units. The Physiography classes plains,undulating land, are observed in the study area. Most of the study area is covered by low category (97 % of the study area). Small part of the study area comes under medium category (3%). (CPCB Technical Guidelines, 1995).

Ground Water Quality Variation in the Study Area

Field work was conducted in the study area from where 99 ground water samples were collected at different locations. The water samples were analyzed for various physico-chemical parameters. The results obtained are presented in Table 1 and each parameter spatial distribution maps and salt water intrusion zones map (Figure 4) prepared , based on this the effected villages are identified in the study area.

pH

pH of groundwater samples was ranging from 6.9-8.2 which shows the alkaline nature of water. The pH is an important variable in water quality assessment as it influences many biological and chemical processes with in a water body and all processes associated with water supply and treatment.

Electrical Conductivity

EC of groundwater ranged from 293-7380 mg/l. The high concentration of EC may be due to high concentration of ionic constituents present in water bodies and reflects contributions from salinity intrusions as well as Aquacultute effluents.

Total Dissolved Solids

The TDS values of groundwater sample ranged from 190-4797 mg/l. The observed TDS values in most of the water samples are exceeding the permissible limits of ICMR. The suspended solids values indicated the presence of organic and inorganic solids that can provide adsorptive sites for certain chemicals and biological agents. The high TDS concentrations are observed in near to the brackish water and salt ponds villages.

Hardness

The hardness in the groundwater samples ranged from 4-1180 mg/l. The concentration of hardness is above the prescribed limit of 300 ppm. in most of the samples collected from study area. Hard water is high in dissolved minerals, both calcium and magnesium. As water moves through soil and rock, it dissolves small amounts of these naturally-occurring minerals and carries them into the ground water supply. Water is a great solvent for calcium and magnesium, so if the minerals are present in the soil around well and its water supply, can end up with hard water EPA (1996).

The hardness values ranged from 80-1140 mg/l. Areas such as Nagarau Topu,showed a concentration > 1140 mg/l. Hence these waters can be considered as very hard water.

Alkalinity

The alkalinity values of groundwater samples ranged from 160-870 mg/l, respectively. Large amount of alkalinity imparts a bitter taste to water and usually unpalatable and consumer acceptance decreases with this water.

Chlorides

Chlorides values of groundwater samples ranged from 35-91402 mg/l, respectively which is above the prescribed standards NEERI (1990) in few samples. Water containing chloride in excess of 250 mg/l are considered to be undesirable for drinking purposes. Sodium chloride exerts a salty taste to water and excess chloride causes cardio vascular diseases. The chloride concentration in ground water of the study area varied between 30 mg/l-270 mg/lt. Chloride concentrations of the study area exceed the permissible limit in the several areas.

Sodium

The permissible limit for sodium as per Bureau of Indian Standards is 200 ppm.s. The value present in the study area varies from 37-927 mg/l. 42-1227 mg/l. Agricultural drainage water was the main cause to increase the sodium consentration levels in the study area.

Potassium

The major natural source of potassium in ground water is Feldsparn, some Micas and clay minerals. Potassium is corrosive to metals. The permissible limit for potassium as per Bureau of Indian Standards is 200 ppm. Potassium values obtained showed no variation for the present study area and they are considered with in the permissible limits. The value present in the study area varies from 1-144 mg/l.

Calcium

The calcium values of groundwater samples ranged from 6-740 mg/l, respectively.

Magnisium

The Magnisium values of groundwater samples ranged from 6-740 mg/l, respectively.

Saline Water Intrusion Zones

The present study shows that the lower Pennar basin has severely effected by the Saline water intrusion. Delineation of saline water areas based on the TDS (mg/l). Table 2 shows water sample analysis data Through out study with various parameters Salinity map was generated by the spatial distribution of TDS (mg/l).This study area was fell in 3 categories i.e., Fresh, Slightly saline, Moderately Saline. Very saline water observed in Brackish water and salt ponds ranging from 10,000 to 26,000 mg/l and above, Moderately saline observed near by Brackish water and poor drainage areas ranging from 3000 above to 6000 mg/l. saline waters were observed in aqua culture areas, and Fresh water zones were observed in reaming parts of study area.

CONCLUSIONS AND RECOMMENDATIONS

From the results obtained from the above study aquaculture and salt cultivation impact on ground water quality effected village wise saline zones are identified the effected villages are Kancherla Palem, Palle Palem, Patti Palem, Utukuru, Alaganipadu, Vavilla, Gadiladinne, Kura Palem, Veerareddy Palem, Ramudu Palem, Narsapuram, Venkahna Palem, Kudithi Palem, Indupooru, Purini, Pedda Palem, Dundigunta, AlluraMaipadu, Ganga Patnam. Based on the physico-chemical analysis of ground water and the water quality index, the areas under impact i.e., areas where high concentration of water quality parameters is observed. The results indicate that certain parameters like nitrates, TDS, hardness and alkalinity exceeded the permissible limit as given by Bureau of Indian Standards. The concentration of Sodium, Potassium, Chloride, Sulphates however was within permissible limits. The poor quality which all lie in the residential and Agricultural and salt cultivation areas of the study area.

Recommendations

The following recommendations are suggested to improve the prevailing environmental crises in the study area

- A detailed salinity - discharge velocity study should be carried out for the different times periods and for the different rates of fresh water outflow during selected tidal cycles in order to satisfactory represents the salinity - river discharge relation ship in the study area.
- To determine the saline - fresh water interface and the degree of both surface and ground water contamination by saline water intrusion. More ground and surface water monitoring stations should be set up on a grid basis. Moreover, geophysical surveys should be performed for comparison with hydro chemical studies.
- The construction of coastal embankments and polder development that is planned to considered the long-

term environmental problems/consequences to improve water quality in shallow aquifers by allowing freshwater recharge that flushes out saline water from the aquifer.

- Emphasizing more salinity-tolerant crops so as to get maximum output should modify the cropping pattern in an area.
- Unavoidable drainage waters should be intercepted, isolated, and reused to irrigate a succession of crops of increasing salt tolerance, possible including eucalyptus and halophyte species, so as to reduce drainage water volumes further and to conserve water and minimize pollution, while producing useful biomass.
- Co-ordination among geoscientists, engineers, environmentalists and planners/policy makers is important for the whole spectrum to be a multidisciplinary approach.

REFERENCES

Karanth K. R. (1987), Ground Water Assessment, *Development and Management*, 667-684.

Ramesh R., and M. Anbu (1996), Chemical Methods for Environmental Analysis, Water and Sediment.

Hassan M. Q. (1992), Saline Water Intrusion and Hydrological Modeling in South West Bangladesh. Schelzky and Jeep Publ, Berlin, 157.

David Keith Todd (1980), Ground Water Hydrology, John Wiley and Sons.

Thomas M. Lillesand, and Ralph W. Keifer (2000), Remote Sensing and Image Interpretation, John Wiley and Sons, New York.

Central Ground Water Board Manual on Water and Waste Water Analysis (1986), NEERI, Nagapur.

Rama Krishnaiah K., The Story of Pennar Basin, Andhra Pradesh, Vani Printing Works, Nellore.

Thomos M. Stein (1997), Sea Water Intrusion in Coastal Aquifers Guidelines for Study, *Monitoring and Control.*, Water Report-11, U.S.A.

Ghazall A. M. (1997), Modelling of Contaminant Transport and Seawater Intrusion in Coastal Aquifers, *www.wiz.unikassel.de*

Integrated Remote Sensing and Factor Analytic GIS Model for Evaluating Ground Water Pollution Potential, O. P. Dubey, D. C. Sharma from Irrigation Research Institute, Roorkee, India.

14

ANALYSIS OF METROLOGICAL DROUGHT USING GIS

A. S. Chandra Bose and G.K.Viswanadh

INTRODUCTION

Drought is one of the major natural environmental hazards and has grave socio- economic impacts. It severely affects the water resources and vegetation. Drought indicates the deficit of rainfall, runoff, water availability, soil moisture, ground water, surface water and crop yield that have been the subject of many investigators from the previous century. The National Commission on Agriculture (1976) broadly classified droughts into Meteorological drought, hydrological drought and agricultural drought. Meteorological drought is a situation when the actual rainfall is significantly lower than the climatologically expected rainfall over a wide area. Hydrological drought is associated with marked depletion of surface water causing very low stream flow and consequent drying up of lakes, rivers, reservoirs etc. Hydrological drought results if meteorological drought is sufficiently prolonged. Agricultural drought is a condition in which there is insufficient soil moisture available to a crop resulting in reduction of yield. Agricultural drought is the end effect of meteorological and hydrological drought and its severity is assessed from the crop yield. So, the periodical drought assessment in a region is essential to take up the management measures (Giridhar and Viswanadh).

Many researchers related agricultural drought by an inadequate amount of soil water available during the critical crop growth periods. Subramanyam (1967) and Subramanyam *et al.*, (1974) considered as a shortage of moisture estimated through simple water budgeting procedure. The index developed for drought assessment was known as Moisture Adequacy Index (MAI), which is ratio of water deficit to over water need. Palmer (1986) defined severity of agricultural drought in terms of the magnitude of computed abnormal evapotranspiration deficit and expressed as a crop moisture index. The model parameters used in water budgeting procedure was developed based on the longer records of hydro meteorological data. The objective of the present study is to assess meteorological drought and to prepare thematic maps using Arc GIS 9.2

STUDY AREA

The Godavari basin is situated between East longitudes 73° 21' to 81° 09' and North latitudes 16° 07' to 22°50' in the Deccan plateau covering large areas in the States of Maharashtra, Madhya Pradesh, Chattisgarh, Orissa, Karnataka and Andhra Pradesh. The Godavari basin extends over an area of 3, 13, 812 sq.km which is nearly 10% of the total geographical area of the country. Godavari catchment was divided in to eight sub basins. In the present study, G-5 sub basin i.e. 'Middle Godavari' sub basin of River Godavari has been considered to analyze meteorological drought at each rain gauge sub area. The Pranahita sub basin lies between latitudes 17°04' and 18°30' N and longitudes 77°43' and 79°53' E. The 'Middle Godavari' sub basin has a catchment area of 35723 km^2, which constitutes 11.38% of the total basin. The catchment area of the 'Middle Godavari' sub basin entirely lies in the State of Andhra Pradesh. In the present study, Middle Godavari sub basin has been considered up to Kaddam reservoir only which lies between latitudes 19°05' and 19°35' North and longitudes 78°10' and 78°55' East. The areal extent of the study area is 2651 km^2, which constitutes 7.4% of the sub basin area. Eight rain gauge stations Khanapur, Kaddam, Neredigonda, Boath, Ichchora, Utnoor, Inderavelli, Bazarhatnoor, were identified in and around the study area. Daily rainfall data was collected for all ten rain gauge stations in the study area for a period of 1996 to 2006 from Bureau of Economics and Statistics, Khairathabad, Hyderabad.

DATA & METHODOLOGY

Software Used

- Arc GIS 9.2
- Micro soft Excel

Data Used

- Daily rainfall data from 1996 to 2006
- Toposheets used : 56-I 3, I 6, I 7, I 8, I 10, I 11, I 12, I 14, I 15 &I 16

IMD Meteorological Drought Assessment

Thiessen network was developed using Arc GIS 9.2 software. Normal rainfall has been calculated at Khanapur, Kaddam, Neredigonda, Boath, Ichchora, Utnoor, Inderavelli, Bazarhatnoor as 881.12 mm, 990.89 mm, 822.27 mm, 1031.96 mm, 1041.99 mm, 1089.66 mm, 1210.49 mm, and 946.97 mm respectively. In the sub basin, highest rainfall was recorded at Inderavelli rain gauge station as 1210.49 mm and lowest is 822.27 mm at Neredigonda rain gauge station.

The IMD uses Irrigation Commission report, 1982 to assess the meteorological drought on the basis of percentage deviation of rainfall from the mean rainfall. The percentage deviation is given by the following equation (1).

$$MD_t = \frac{P_{\text{act}} - P_{\text{mean}}}{P_{\text{mean}}} X100 \qquad (1)$$

where

MD_t = Percentage of meteorological deviation,

P_{act} = Actual rainfall in the time period 't',

P_{mean} = Long term mean rainfall.

The percentage deviation of rainfall and the category of drought assessment used by IMD are as given in the Table 1. The IMD drought assessment method was used for Eight rain gauge stations in G-5 sub basin of river Godavari. Eleven water year annual rainfall data average and month wise percentage deviations were calculated. Corresponding month wise drought severity values for eight rain gauge stations are presented in Table 2 and Table 3.

Table 1

Percentage of Deviation and Category of Drought Assessment

Sl. No	Percentage Deviation (MD_t)	Class	Category
1	>0	M0	No drought
2	0 to -25	M1	Mild
3	-25 to -50	M2	Moderate
4	<-50	M3	Severe

*Source:*IMD Drought assessment/Irrigation commission report 1982.

RESULTS AND DISCUSSION

Based on the results obtained from the IMD drought assessment method for the study area, the drought severity estimates were arrived and discussed below. As seen from the tables, it is observed that all the mandals had experienced severe drought in the months of January and February for the year 2006.There was no drought in the month of September in all the mandals except Boath mandal. Similarly, there was no drought in the month of March in all the mandals except Inderavelli. No drought was experienced even in the month of April except Boath and Kaddam. In the month of August also, there was no drought except in Indervelly and Neredigonda. Mild to moderate drought was experienced in the month of October in all the mandals except in Boath and Indervelly. In the month of November, severe to No Drought was experienced. There was no drought in the month of December in all the mandals except Boath. Boath, Ichchoda

Table 2
Yearly Rainfall Distribution at Each Rain Gauge Station

Year	*Boath*	*Bazarhat-noor*	*Ichchoda*	*Indervelly*	*Kaddam*	*Khanapur*	*Neredigonda*	*Utnoor*
1996	1060.8	1071	684.5	805.3	862.1	857.8	541.4	966.4
1997	789.4	747.7	760.3	1060.4	831.4	791.2	587.8	789.8
1998	1024.1	1186.6	1051.8	1151.2	1144.1	872.9	938.8	889
1999	1128.5	149	1162.3	1317.8	918.2	786.6	178.1	1154.6
2000	983.1	963.8	1114.5	1317.2	1071.1	733.8	1006.7	1327.4
2001	1051.9	1146.3	1240.6	1079.5	1016.6	758.9	1148.3	1122.7
2002	1052.3	1080.5	1099.4	1265.4	909.7	743.8	1091.1	1114
2003	1389.7	1400.1	1427.4	1719.6	1301.7	1046.3	1083.3	1294.2
2004	603.1	619.5	585.3	969.2	734.4	725	565	844.4
2005	1194.4	1105.2	1293.8	1405.1	1119.6	1494.9	1082.2	1394.1
2006	1012.7	1289.1	1174.1	1374.9	1081.9	1092.2	1040	1413.3
Normalrainfall	1031.96	946.97	1041.99	1210.49	990.89	881.12	822.27	1089.66

Table 3
Illustration of Drought Severity – IMD for the year 2006

Year	*Boath*	*Bazarhat-noor*	*Ichchoda*	*Indervelly*	*Kaddam*	*Khanapur*	*Neredigonda*	*Utnoor*
Jan	-100.0	-100.00	-100.00	-100.0	-100.00	-100.00	-100.00	-100
Feb	-100.0	-100.00	-100.00	-100.0	-100.00	-100.00	-100.00	-100
Mar	1514.3	271.68	279.16	-100.0	766.44	811.65	1156.41	87.29
Apr	-100.0	33.66	12.54	221.5	-14.39	304.48	172.73	149.13
May	99.6	651.43	126.85	-13.5	-63.52	-13.17	124.95	-56.13
June	-24.2	11.10	-19.70	-87.8	-40.31	-19.99	-12.86	-48.07
July	-11.4	0.44	-9.79	-32.5	-30.31	-12.61	26.94	2.90
Aug	3.7	49.26	19.05	-23.1	29.36	15.05	-1.95	57.34
Sept	32.6	104.38	86.60	249.4	119.89	142.11	194.77	170.63
Oct	-55.3	-45.96	-33.40	299.1	-24.15	-26.22	-42.10	-7.62
Nov	-34.1	62.79	73.91	1566.7	-100.00	-100.00	-100.00	388.57
Dec	-100.0	0.00	0.00	0.0	0.00	0.00	0.00	0.00

Table 4
Monthly Drought Severity Indices in the Study Area for the year 2006

Year	*Boath*	*Bazarhat-noor*	*Ichchoda*	*Indervelly*	*Kaddam*	*Khanapur*	*Neredigonda*	*Utnoor*
Jan	Severe	Severe	Severe	Severe	Severe	Severe	Severe	Severe
Feb	Severe	Severe	Severe	Severe	Severe	Severe	Severe	Severe
Mar	No drought	No drought	No drought	Severe	No drought	No drought	No drought	No drought
Apr	Severe	No drought	No drought	No drought	Mild	No drought	No drought	No drought
May	No drought	No drought	No drought	Mild	Severe	Mild	No drought	Severe
June	Mild	No drought	Mild	Severe	Moderate	Mild	Mild	Moderate
July	Mild	No drought	Mild	Moderate	Moderate	Mild	No drought	No drought
Aug	No drought	No drought	No drought	Mild	No drought	No drought	Mild	No drought
Sept	No drought	No drought	No drought	No drought	No drought	No drought	No drought	No drought
Oct	Severe	Moderate	Moderate	No drought	Mild	Moderate	Moderate	Mild
Nov	Moderate	No drought	No drought	No drought	Severe	Severe	Severe	No drought
Dec	Severe	No drought	No drought	No drought	No drought	No drought	No drought	No drought

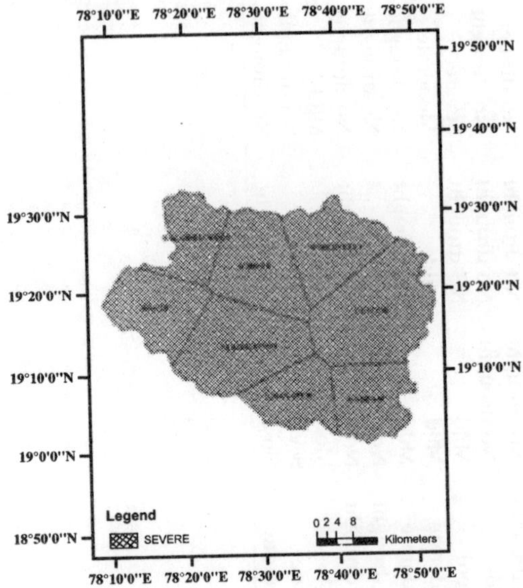

Figure 1 Monthly Drought Assessment – January 2006

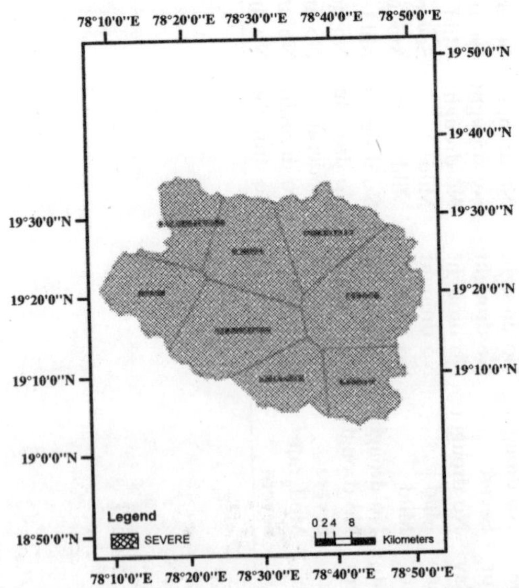

Figure 2: Monthly Drought Assessment – February 2006

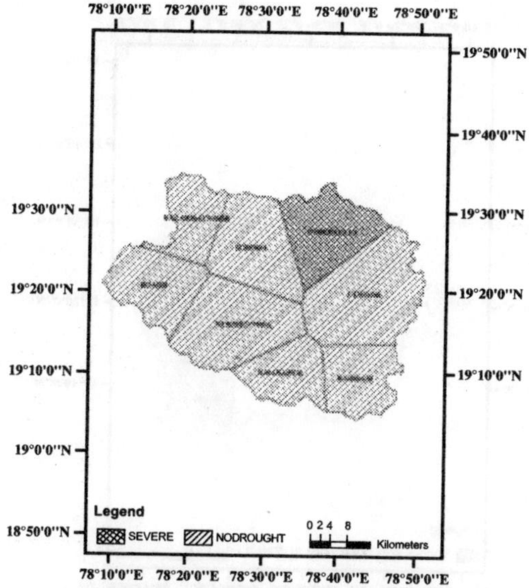

Figure 3: Monthly Drought Assessment – March 2006

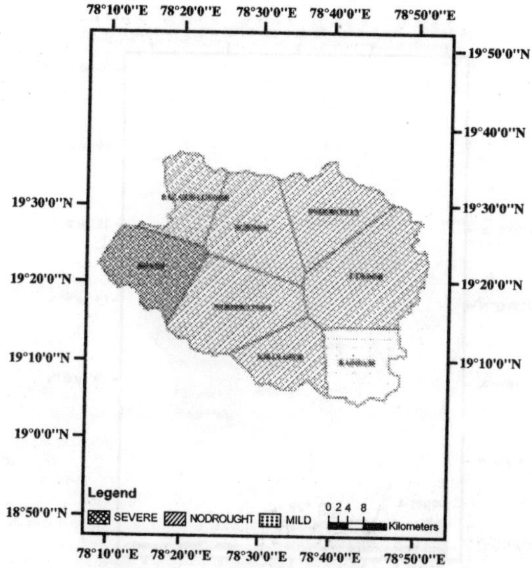

Figure 4: Monthly Drought Assessment – April 2006

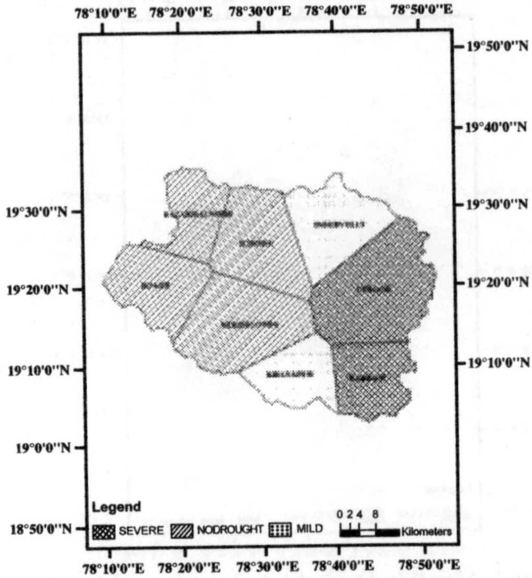

Figure 5: Monthly Drought Assessment – May 2006

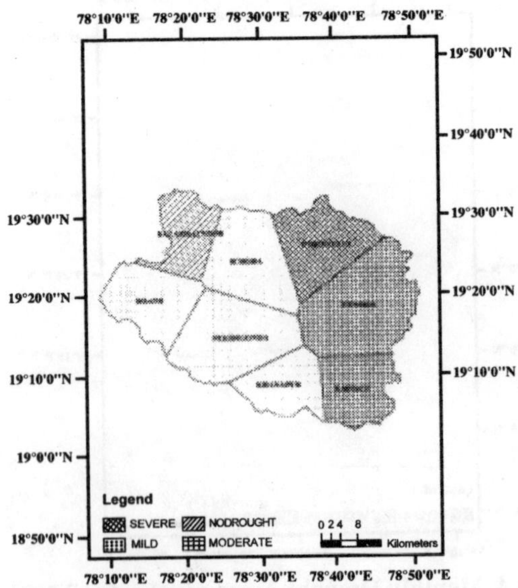

Figure 6: Monthly Drought Assessment – June 2006

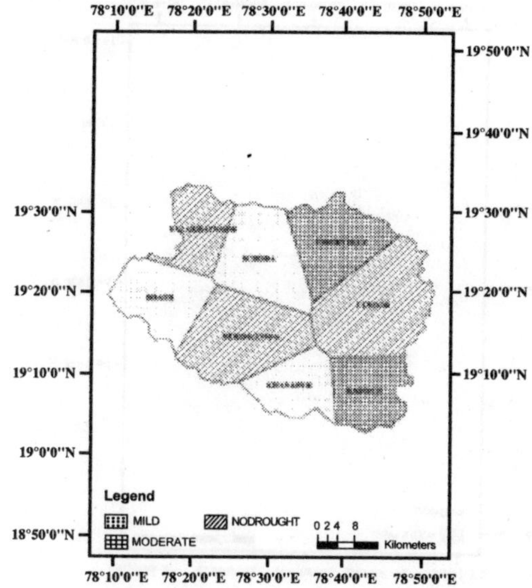

Figure 7: Monthly Drought Assessment – July 2006

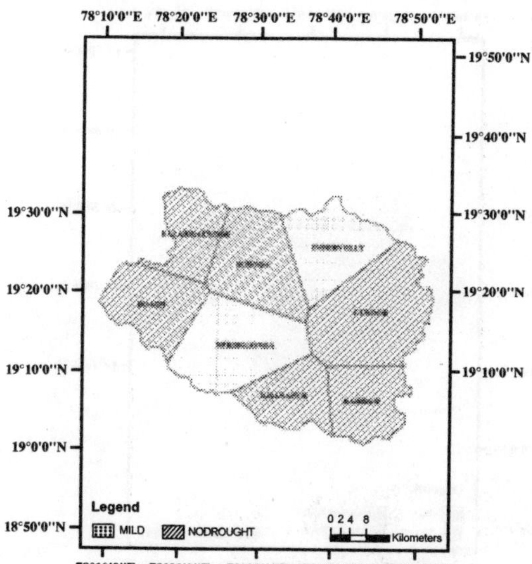

Figure 8: Monthly Drought Assessment – August 2006

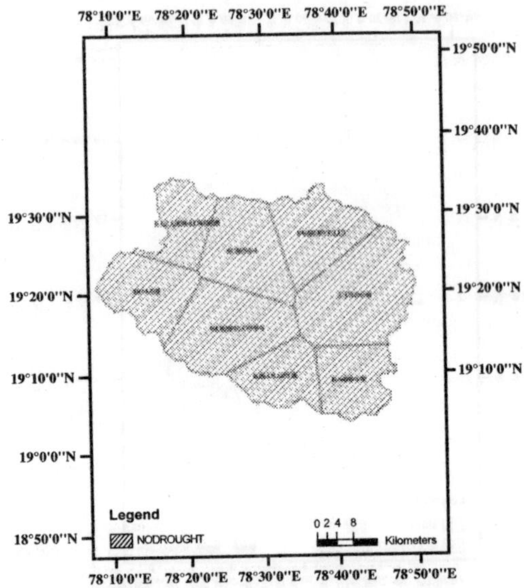

Figure 9: Monthly Drought Assessment – Septemver 2006

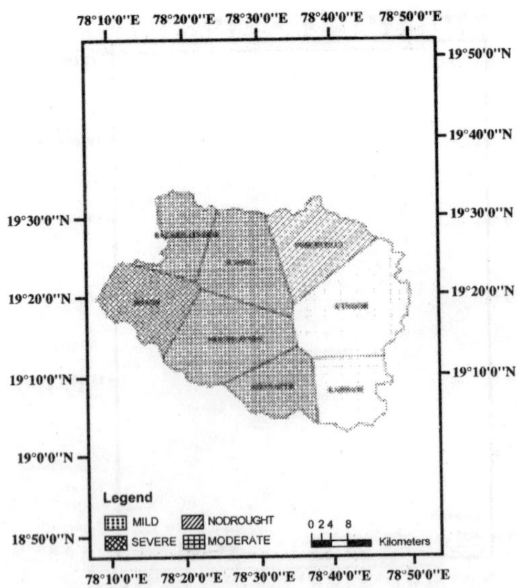

Figure 10: Monthly Drought Assessment – October 2006

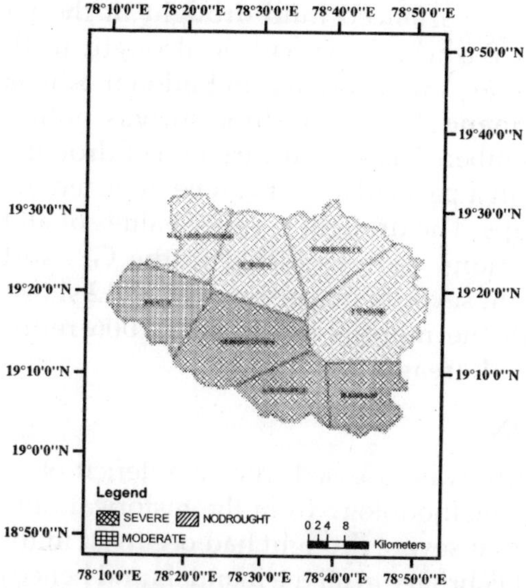

Figure 11: Monthly Drought Assessment – November 2006

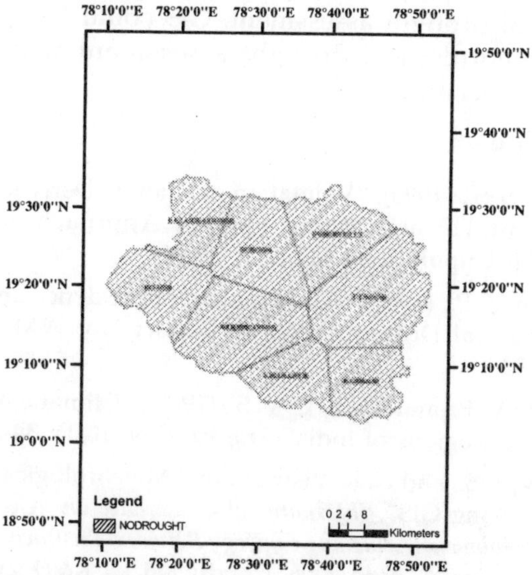

Figure 12: Monthly Drought Assessment – December 2006

and Khanapur experienced mild drought in the months of June and July. Bazarhatnoor had no drought in the entire year except January and February. In Kaddam, Khanapur and Neredigonda mandals, severe drought was noticed in the month of November. The spatial variation of drought severity over the basin in a particular year can be depicted in the form of thematic maps. The drought severity values of all the eight rain gauge stations were exported to the GIS software to generate drought severity maps. Figures1 to 12 show drought severity for all the months in the year 2006 represent the drought affected areas.

CONCLUSION

Drought severity was assessed from the deficit of rainfall by using the IMD methodology from the historical rainfall data. It is observed that severe drought had occurred in the months of January and February in all the mandals and hence judicious use of Ground water in months as well as other precautionary measures are warranted. To know the realistic actual drought situation in a basin, hydrological drought and agricultural drought assessment must be taken up along with the meteorological drought assessment. GIS could be effectively used for meteorological drought assessment with greater efficiency and accuracy.

REFERENCES

Giridhar M.V. S. S., (1995), "Validation of Satellite Derived Drought Assessment Through Water Balance Approach at Mandal Level," An Unpublished M. Tech Thesis.

Subramanyam V. P., (1967), "The Problem of Incidence and Spread of Continental Drought", WMO Report No: WMO/IHD-2, Geneva: 52.

Subramanyam V. P., and Sastry C. V. S., (1969), "Climatic Analogues for the Dry Regions of India" *Tropical Ecol.* 10(1): 34-54.

Giridhar M. V. S. S., and G. K. Viswanadh, "Meteorological Drought Analysis Using GIS", *International Conference on "Geomatics for Rural Development-Achieving Synergy Between Technical and Social Systems"*, 5th to 8th June 2006, Conducted by R&D Cell, *JNTU*, Hyderabad, India.

15

AN INTEGRATED APPROACH FOR GROUND WATER RESOURCES MAPPING USING REMOTE SENSING & GEOGRAPHICAL INFORMATION SYSTEM

S. S. Asadi and S. Srinivasulu

INTRODUCTION

Water is a vital and a key resource in all-economic activities ranging from agriculture to industry. Only a tiny fraction of the planet's abundant water is available to us as fresh water (Hassan *et al.*, 1974). About 97% is found in the oceans and is too salty for drinking, irrigation, or industry. The remaining 3% is fresh water. About 2.997% of it is locked up in ice caps or glaciers or is buried so deep that it costs too much to extract. Only about 0.0035 of Earth's total volume of water is available. In India of total available water resource 1869 bcm, the usable, water resources are only 1122 bcm, which consists surface water 690 bcm, ground water 432 bcm, which the present per capita available water resources is 1122 cm and by 2050 it is likely to reduce to 748 cm. When the countries per capita water availability is less than 1700 cm it is considered as water stress country (Seetharaman, 2004).

In present days, usage of surface and groundwater is increasing due to rapid urbanization, industrial growth and agricultural utilizations. Consequently, rapid depletion of surface and ground water is taking place (Anand Kumar *et al.*, 2002). The conventional methods like geological and geophysical studies are generally used to locate bore well points. Compare to the conventional methods Remote Sensing is the advanced technology, because of its wide coverage and repetitivity, to identify sub-surface water resources (Shah, 2001). Keeping this in view, the present study aims at groundwater prospects mapping to increase groundwater levels at the arid and semi-arid area using Remote Sensing and GIS.

DESCRIPTION OF STUDY AREA

The study area is situated in a part of Mahbubnagar, Hyderabad districts of A.P., bounded by 78°15′ & 78°30′E longitude and 17°0′ & 17° 15′ N latitude falling in SOI toposheet no. 56k/8 falling in SOI toposheet no.56k/8.The Mahbubnagar district part the mean minimum temperature is 13° C to 17° C in December and January, but it rises to 26° C to 29° C in March. The average annual rainfall is 884.8mm.Hyderabad district part is situated is the hottest part of the State during summer and coldest in winter. The mean daily temperature varies from 30° C to 36° C from April to June and from 20° C to 24° C in the months of December and January. The mean maximum temperature ranges between 40° C and 47° C in May (Census of India, 2001). The Hyderabad district area stands on pink and Grey granites. The soils are mainly brownish sandy and Red loamy. Rocks like dolomite and quartz basalt associated with granites yield in general brownish sandy soil and where the dolomite is predominant, the soil changes to black cotton.

STUDY OBJECTIVES

1. To prepare thematic maps from survey of India toposheets and satellite imagery using visual interpretation technique.

2. To collect collateral data from different departments and to create attribute information of thematic maps using GIS tool.
3. To prepare a final ground water prospects map by integrating the lithological, structural, geomorphological and hydrological information (integration of spatial and attribute data).
4. To identify sites for construction of recharge structures to increase the groundwater level based on the ground water prospect map.

METHODOLOGY

Data Collection

Different data products required for the study include Survey of India (SOI) toposheet (56K/8), fused data of IRS-1D PAN and LISS-III satellite imagery obtained from National Remote Sensing Agency (NRSA) and collateral data from related organizations comprising of collateral data and demography data.

Database Creation

Satellite imageries are georeferenced using the ground control points with SOI toposheets as a reference and further merged to obtain a fused, high resolution (5.8 m of PAN) and colored (R, G, B bands of LISS-III) output in EASI/PACE Image processing software. The study area is then delineated and subsetted from the fused data based on the latitude and longitude values and a final hard copy output is prepared for the generation of thematic maps using visual interpretation technique. These thematic maps (raster data) are converted to vector format by scanning using an A0 flatbed deskjet scanner and digitized using AutoCAD software for generation of digital thematic maps using Arc/Info and ARCVIEW GIS software. The GIS digital database consists of thematic maps like land use/land cover, drainage, soil, physiography, slope, road network, geomorphology, structures, ground water prospects using Survey of India (SOI) toposheets and fused data of IRS-ID PAN and IRS-ID LISS-III satellite imagery.

Spatial Database

Creating a GIS spatial database is a complex operation, which involves data capture, verification and structuring processes. Raw geographical data are available in many different analogue and digital forms such as toposheets, aerial photographs, satellite imageries and tables. Out of all these sources, the source of toposheet is of much concern to natural resource scientist and an environmentalist (Madan Mohan, 2001). In the present study, the layers generated from toposheet and satellite data using visual interpretation technique include base map, drainage map, and slope map prepared using road network, drainage pattern, contours and spot heights from SOI Toposheet. The procedure consists of a set of image elements, which help in the recognition or interpretation of various land use/land cover features systematically on the enhanced satellite imagery during the classification of features (Thomas M. Lillesand, 2000). The thematic maps generation system used in this study is the system, which is pioneered by United States Geological Survey (USGS) (Anderson, 1977) and is modified by National Remote Sensing Agency (NRSA) according to Indian conditions. A preliminary image classification key is prepared for the fused pictorial data and is used during interpretation process. The base map is overlaid on the satellite imagery. Then the different thematic layers features are extracted from the satellite pictorial data. The doubtful areas (due to similar spectral response and spectral signature) identified during the preliminary image classification are listed out before ground verification. The doubtful areas are physically verified by field observation about terrain condition and land use pattern. Based on the ground information collected, corrections and modifications of mis-classified details and doubtful areas are carried out on enhanced imageries for extracting final thematic maps (National Remote Sensing Agency, 2000) so as to extract the entropy or information extent. At the end of the interpretation process the above thematic maps in the form of paper-based maps.

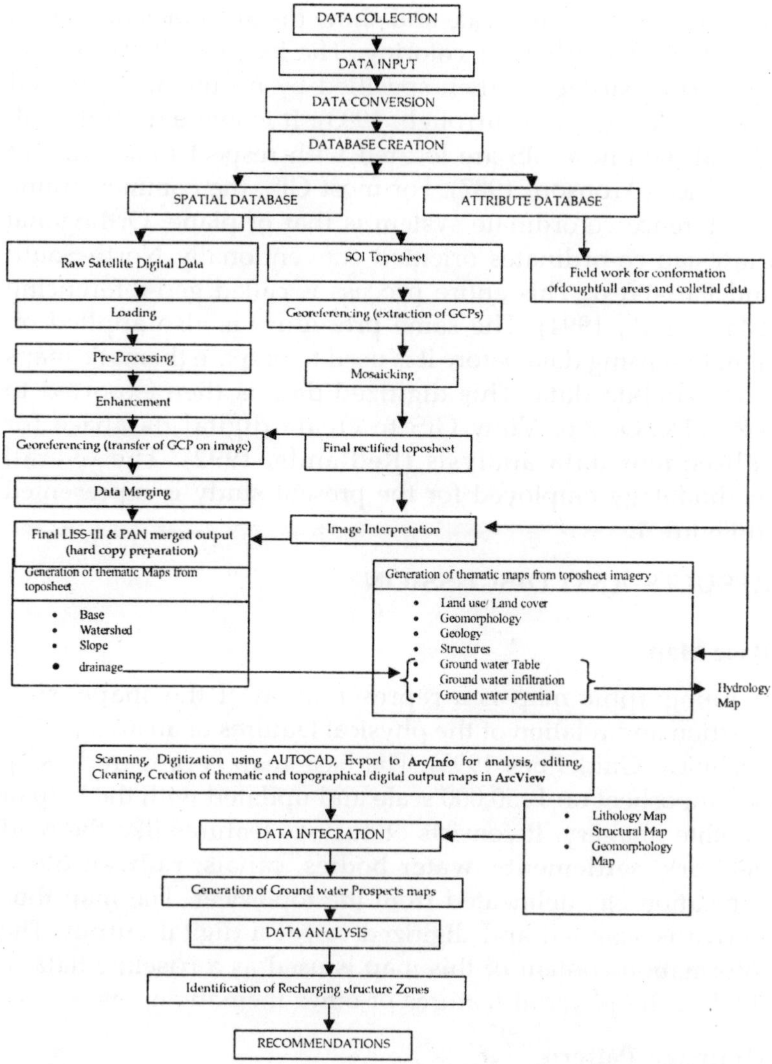

Figure 1: Flow Chart Showing the Methodology Adopted for the Present Study

Attribute Database

The thematic maps are converted to digital mode using scanning and automated digitization process. These maps are prepared to a certain scale and show the attributes of entities by different symbols or coloring. The location of entities on the earth's surface is then specified by means of an agreed co-ordinate system (Burrough, 1983). It is mandatory that all spatial data in a GIS are located with respect to a frame of reference (Aronson, 1987). For most GIS, the common frame of reference co-ordinate system is that of plane, Orthogonal Cartesian co-ordinates oriented conventionally North-South and East-West. This entire process is called geo-referencing (Mark *et al.*, 1994). The same procedure is also applied on remote sensing data before it is used to prepare thematic maps from satellite data. This digitized data is then exported to ARC/INFO, Arc View GIS to create digital database for subsequent data analysis (Redlands, 1992). The overall methodology employed for the present study is represented in Figure 1.

RESULTS AND DISCUSSION

Base Map

A topographic map is a representation of the shape, size, position and relation of the physical features of an area (IMSD Technical Guidelines 1995). The base map is prepared using SOI toposheet on 1:50,000 scale and updated with the help of satellite imagery. It consists of various features like the road network, settlements, water bodies, canals, railway track, vegetation etc. delineated from the toposheet. The map thus drawn is scanned and digitized to get a digital output. The information content of this map is used as a baseline data to finalize the physical features of other thematic maps.

Drainage Pattern

The drainage map prepared from the toposheet forms the base map for the preparation of thematic maps related to surface and groundwater. All the rivers, tributaries and small stream channels shown on the toposheet are extracted to prepare the drainage map. Care is taken that the boundaries of rivers/water bodies appearing on land use/land cover map or base map are perfectly matched with those on the toposheet. All the drainage lines are examined very closely and final drainage map is prepared (Thomas M. Lillesand *et al.*, 2000)

Watershed Characteristics

Watersheds are hydrological units that are considered to be efficient and appropriate for assessment of available resources and subsequent planning and implementation of various development programmes. The watershed map is prepared in accordance with the National Watershed Atlas and River Basin Atlas of India, 1985. According to this, India is divided into 6 regions out (River Basin Atlas of India, 1985) of which the present study area comes under Region-4 and part of basins D & E. The study area is under basin D (4D1E5, 4D1E6, 4D1E8, 4D2B7).

Slope Map

Slope classes 1, 2 and 3 are observed in the study area. Most of the study area is covered by nearly level slope class (92% of the study area). Small part of the study area (4%) comes under very gently sloping class and 2% study area comes under the moderately sloping (IMSD Technical Guidelines, 1995).

Land Use/Land Cover

The land use/land cover categories such as built-up land, agriculture, forest, water body and wastelands have been identified and mapped from the study area (Figure 2). Major part of the study area is covered with single crop (94%). About 0.013% of the study area is under built-up land and Industrial area is 0.016%. From the satellite data the agriculture area (96.02%) could be clearly delineated as four categories, single crop, double crop, fallow land and plantations. Though single crop has been observed at various parts of the study area, double crop has been observed on the west part of the area and plantations are observed at some places of the study area. Water bodies occupied 0.19%. About 0.48% of the study area is under scrub forest and 4.22% of area is under wasteland. Under this category land with scrub (3%), land without scrub (0.22%) and barren sheet rock area (0.09%) are observed. Based on the land use/land cover categories, infiltration rates of water are less at built-up land and barren sheet rock areas. Consequently groundwater potentials at these places are low, whereas the places where water bodies are present have high potentials and moderate potential at remaining categories (Campbell, 1983).

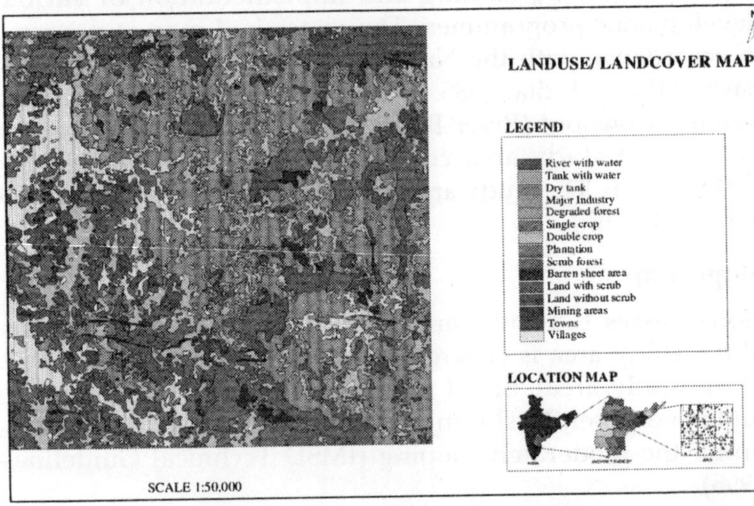

Figure 2: Land Use/Land Cover Map

Geomorphology

The geomorphological classes observed in the study area are pediplain with moderate weathering (PPM) (49%), pediplain with shallow weathering (PPS) (38%), pediment (8%), pediment inselberg complex (PIC) (4%) and dyke (0.12%) (Figure 3). Based on the geomorphology classes the weathering thickness of the soil is 10-12m in PPM, 0-10m in PPS, and in Pediment, PIC classes the soil weathering thickness is very less. Infiltration rates are moderate to good in PPM. Moderate to good yields of groundwater are observed in

PPS, in pediments and PIC classes poor to nil yields are observed. The pediment and pediment inselberg complex are observed as run-off zones and these classes are limited to poor recharge (Schneider *et al.*, 1979).

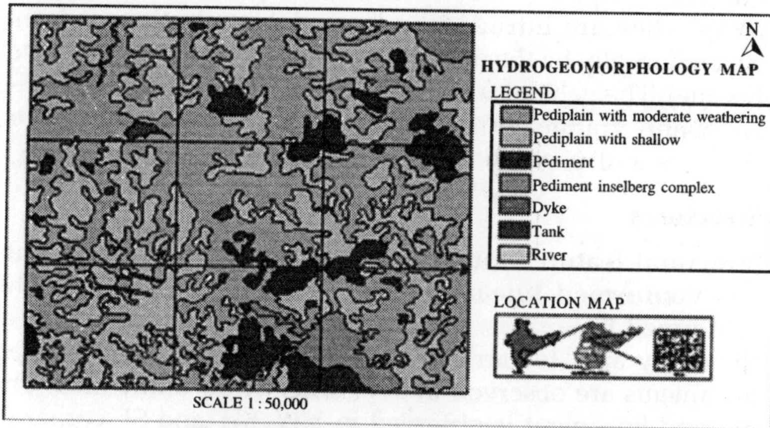

Figure 3: Hydro Geomorphology Map

Geology

The study area is undulating with scattered denudational hills and ridges in the granite gneiss terrain. The altitude varies from 500 mts in the West 400mts in the East with maximum

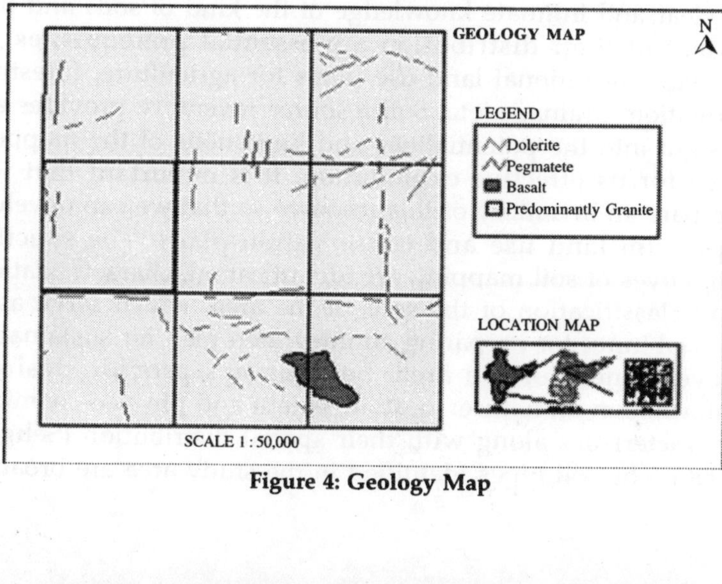

Figure 4: Geology Map

elevation of 650 mts in the SW region 29 (Hyderabad Quadrangle Map, Prepared by Geological Society of India). The study area constitutes mainly a granitic terrain (pink-grey) exposive a variety of archaean granitorides of peniusular gneissic complex (PGC) and schistoic (older metamorphic) rocks. They are intruded by basic dykes (Proterozoic) and covered locally by the deccan traps (upper cretaceous to lower Eocene). The geological categories observed in the study area are mainly granite (99%), basalt (1%), and some of lineaments, dolerites and pegmatites (Figure 4) (Mary Hoffman, 1993).

Structures

Structural features found in the study area are lineaments i.e., conformed lineament and inferred lineaments. The conformed lineaments are observed in the central portion of the study area towards west to east and major conformed lineaments are observed in SW corner of the study area. The inferred lineament is observed in NW, SW and SE corner of the study area. These structural-geological studies commence by deciphering planar discontinuities in the rocks, with a view to understanding their characteristics, discomposition and spatial relations.

Soil

A clear and intimate knowledge of the kind of soils and the extent of their distribution are essential prerequisites in developing rational land-use plans for agriculture, forestry, irrigation, drainage, etc. Soil resource inventory provides an insight into the potentialities and limitations of the mapped area for its effective exploitation. It is important that we prepare an inventory of this resource so that we can develop optimum land use and conservation plans. The specific objectives of soil mapping are identification, characterization and classification of the soils of the area, which serve as a crucial input for preparing an integrated plan for sustainable development of the area. Soil surveys provide desired information on nature, location, extent and physico-chemical characteristics along with their spatial distribution (Sehgal, 1996). The soil types identified in the study area are broadly grouped into clayey soil, gravelly clayey soil, gravelly loamy soil, loamy soil and rock outcrops on undulating lands. These soils within each soil type are further subdivided into soils with low available water capacity (LAWC), medium available water capacity (MAWC), high available water capacity (HAWC) and very high available water capacity (VHAWC) (Figure 5).

Infiltration Rate

The infiltration rate plays an important role in determining the contamination risks of groundwater. The infiltration map is prepared based on primary and secondary permeability of the lithological terrain. Depending on the infiltration rate the study area is divided into high, medium, and low zones as per Central Pollution Control Board (CPCB) guidelines. It is interpreted based on soil types and its thickness and weathering status. Infiltration is low in impermeable, compact and sheet rock areas observed in some portion of the study area and are influenced by urbanization.

Groundwater Table

This map is prepared based on information obtained from the Central and State Groundwater Departments, besides

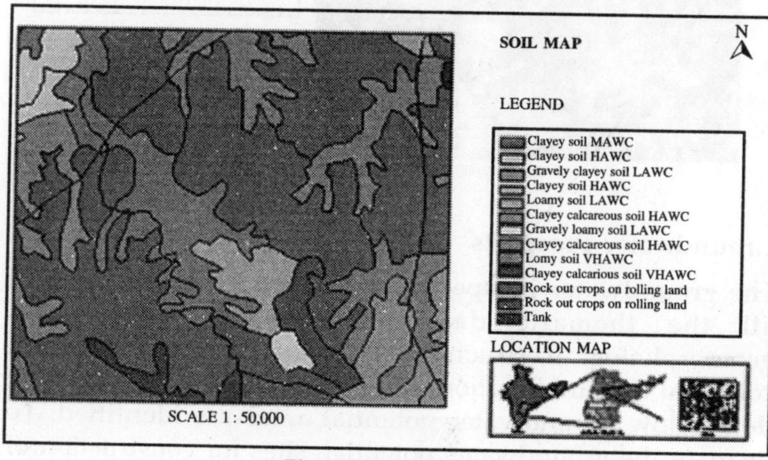

Figure 5: Soil Map

necessary ground checks. During fieldwork, information about the depths of water levels are obtained from the residents and used in preparation of this map.

Groundwater Potential

The groundwater potential map is prepared based on the analysis of various themes such as geomorphology, land use /land cover, lineament, intersection points, drainage pattern, lithological evidences by using converging evidence concept, besides the collateral data obtained from State Groundwater Board with necessary field checks. The groundwater potential map reveals the available quantum of groundwater and is delineated into zones showing excellent, very good to good, good to moderate and poor to nil groundwater potential areas (Figure 6) (Krishnamurthy, 1996).

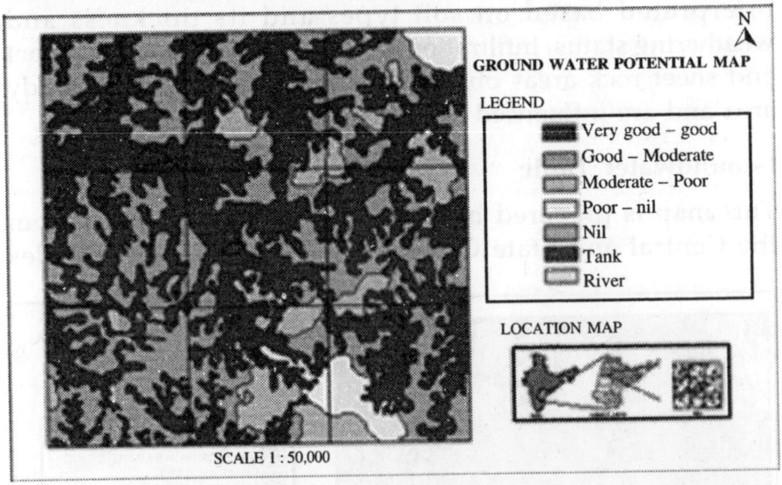

Groundwater Prospects

The groundwater prospects map shows the information of all the thematic data including the geological, geomorphological, structural, hydrological information and collateral data information of the study area. Based on these details low groundwater potential areas are identified. To increase the groundwater potential, sites for construction of

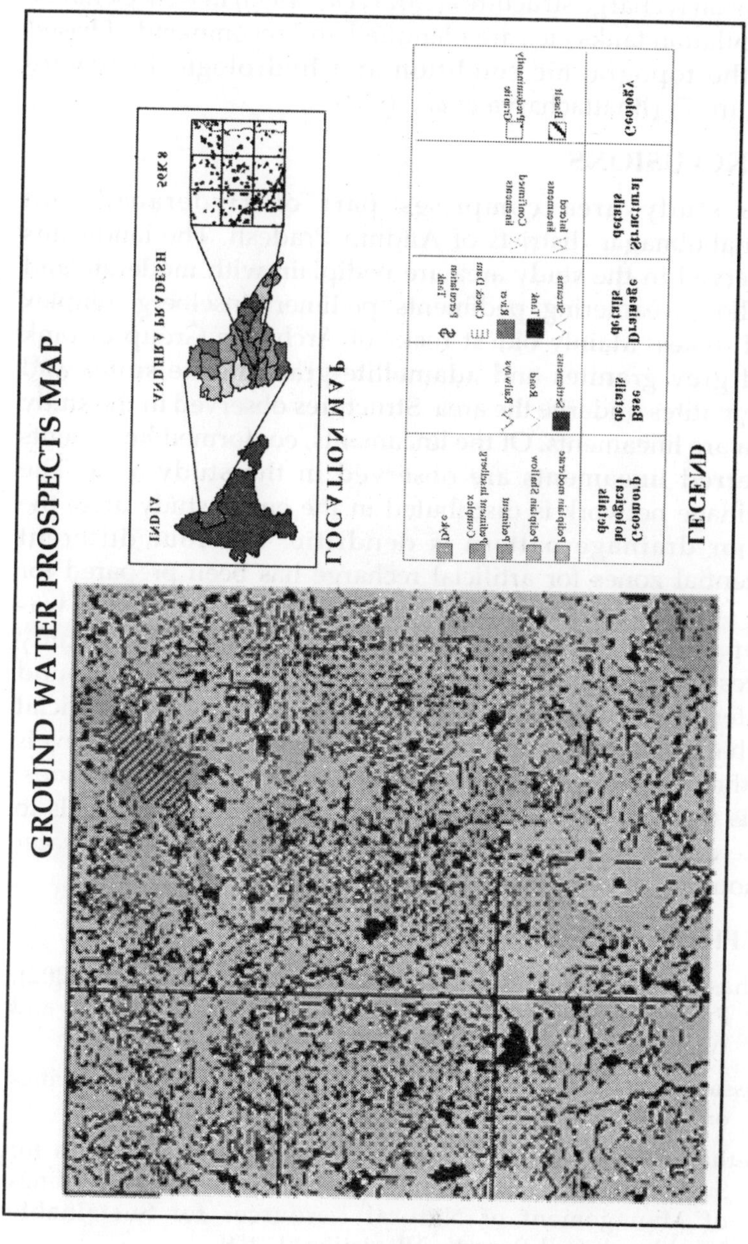

Figure 7: Ground Water Prospects Map

artificial recharge structures (Saraf *et al.*, 1998) like check dams, perculation tanks etc., are identified and recommended based on the topographic condition and hydrological capacity (Figure 7) (Bhattacharya *et al.*, 1973).

CONCLUSIONS

The study area comprises part of Hyderabad and Mahabubnagar districts of Andhra Pradesh. The landforms observed in the study area are pediplain with moderate and shallow weathering, pediments, pediment inselberg complex and dykes. Mainly oldest rocks of Archaean Group of pink and grey granites and adamellite-granodiorite suites and migmatites underlie the area. Structures observed in the study area are lineaments. Of the lineaments, conformed lineaments inferred lineaments are observed in the study area. The drainage network is distributed in the entire study area; the major drainage pattern is dendritic. The four different potential zones for artificial recharge has been prepared for the study area. The final (groundwater prospects) map was prepared by integrating various thematic maps. The study gives the total information of the terrain condition, ground water status and helps identify areas to construct artificial recharge structures in order to improve groundwater levels in the perspective of groundwater use for future generations. This type of study is useful for the administrative officials to take decisions and implement methods for sustainable water resources development.

REFERENCES

Suhas P. Wani, T. J. Rego, P. Pathak, and Piara Singh (2002), Proceedings of International Conference On Hydrology and Watershed Management, l 1: 227.

Hassan A. A. (1974), Water Quality Cycle Reflectance of Activities on Nature and Man, *Groundwater*, 12(1): 16-21.

Seetharaman R. (2004), Management of Natural Resources for Sustainable Livelihoods and Poverty Alleviation, Proceedings of Management of Natural Resources for Sustainable Livelihoods and Poverty Alleviation, 1: 159.

Anand Kumar, and Sanjay Tomar (2002), Application of Remote Sensing and GIS for Groundwater Assessment, *Development Alternatives Newsletter*, 12.

Shah P. N. (2001), Hydrogeomorphological Mapping for The Evaluation of Groundwater Prospect Zones in Mirzapur District, Uttar Pradesh, India Using IRS-1A, LISS-II Geocoded Data, Proceedings of International Conference on Remote sensing and GIS (ICORG), 2: 447.

Census of India (2001), Series-2, Part XII-A & B, District Census Handbook, Ranga Reddy, Published by the Government of Andhra Pradesh.

Burrough P. A. (1983), Geographical Information Systems for Natural Resources Assessment, Oxford University Press, New York.

Madan Mohan (2001), Spatial Data for Environmental Management: A GIS Based Monitoring of Ecological Restoration for Delhi, Proceedings of International Conference on Remote sensing and GIS (ICORG), 2: 546.

Mark D. M., and F. F. Xia (1994), Determining Spatial Relations Between Lines and Regions in ARC/INFO Using The 9-Intersection Model, Proceedings of 14th AnnualESRI User Conference, 1207-1213.

Krishnamurthy J., N. Venkatesa Kumar, V. Jayraman, and M. Manivel (1996), An Approach to Demarcate Groundwater Potential Zones Through Remote Sensing and GIS, *International Journal of Remote Sensing*, 17(10): 1867-1884.

Saraf A. K., and P. R. Chaudhary (1998), Integrated Remote Sensing and GIS for Groundwater Exploration and Identification of Artificial Recharges Sites, *International Journal of Remote Sensing*, 19(10): 1825-1841.

Aronson P. (1987), Attribute Handling for Geographical Information System, *Proceedings of AUTOCARTO 8*, Falls Church, VA: ASPRS, 346-355.

National Watershed Atlas of India (1990), Prepared by all India Soil and Land use Survey, Department of Agriculture and Co-operation, Government of India.

River Basin Atlas of India (1985), Prepared by Central Board for the Prevention and Control of Water Pollution, India.

Technical Guidelines For Disaster Management Information System (DMIS) (1998), Maharashtra Remote Sensing Centre (MRSAC).

Thomas M. Lillesand, and Ralph W. Keifer (2000), Remote Sensing and Image Interpretation, John Wiley and Sons, New York.

Anderson J. R. (1977), Land Use and Land Cover Changes - A Framework for Monitoring, *Journal of Research*, US Geological Survey, 5(3): 143-153.

Campbell J. B. (1983), Mapping the Land: Aerial Imagery for Land Use Information, Resource Publications in Geography, Washington D.C, Association of American Geographers, 96.

Integrated Mission for Sustainable Development (IMSD) Technical Guidelines (1995), Andhra Pradesh State Remote Sensing Application Centre (APSRAC), Government of Andhra Pradesh, December.

Rajiv Gandhi Drinking Water Mission (2000), NRSA Technical Guidelines for Preparation of Ground Water Prospects Maps, *National Remote Sensing Agency.*

Schneider Stanley R., D. F. Mc Ginnis, and John A. Pritchard (1979), Use of Satellite Infrared Data for Geomorphology Studies, *Remote Sensing of Environment*, 8: 313-330.

Hyderabad Quadrangle Map, Prepared by Geological Society of India,

Mary Hoffman (1993), Dictionary of Geology, Special Indian Edition, GOYL Saab Publishers and Distributors, Delhi.

Sehgal J. (1996), Pedology - Concepts and Applications, First Edition, Kalyani Publishers, Ludhiana, India.

Manual for Preparation of Thematic Maps (2002), Industrial Pollution Prevention Project-II (IPPP-II), Consultancy Services for GIS, *Central Pollution Control Board (CPCB).*

16

WOMEN, NATURAL RESOURCES ENVIRONMENT AND LIVELIHOODS DIVERSITY

Siddhartha Sarkar

INTRODUCTION

Poverty and environmental degradation are closely associated and causally interlinked and should therefore be addressed together. The international goal of halving the number of people living in extreme poverty by 2015 and reversing environmental degradation will require addressing rural and urban poverty and environmental degradation simultaneously. At the same time, addressing gender disparities should not be reduced to a means of ensuring the effectiveness of poverty reduction strategies. Gender equality is a development objective in its own right, and sustainable development strategies must aim to foster women's empowerment and effective participation. This implies involving women and men as partners and allies in formulating and pursuing strategies for more equal societies. Over half of the world's poor live in rural areas. Although urban poverty is rising, the correlation between poverty and remoteness from urban centres is strong in most countries and is expected to remain so in the foreseeable future. As compared with their urban counterparts, rural people are often isolated from economic opportunities and have less access to basic social services. Resource degradation is an acute problem in rural areas, with some 60 per cent of the world's poorest people living in ecologically vulnerable areas (Angelsen, 1997). The situation is worst in Africa, with two thirds of the continent being deserts or drylands. Africa also has extensive agricultural dry lands, almost three quarters of which are already degraded (UNCCD, Undated). In many developing countries, declining rates of yield growth and accelerating resource degradation contribute greatly to conflict over natural resources. Food insecurity and malnutrition are critical concerns. Even for land-owning households, farming alone

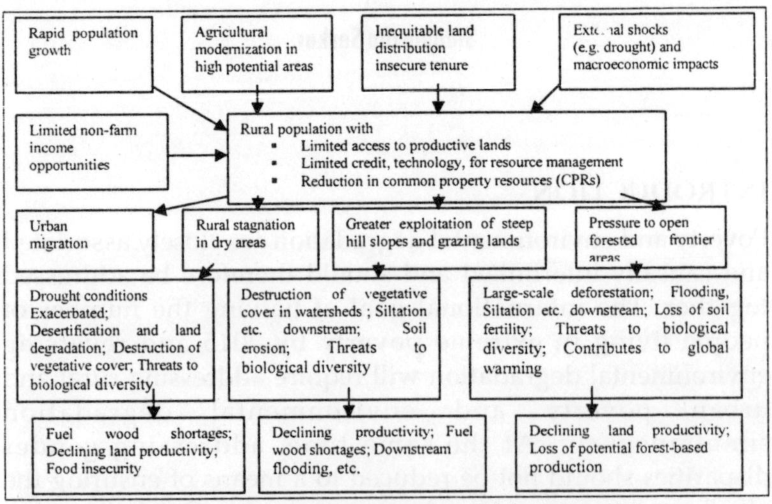

Figure 1: Poverty and Environment Link in Rural Areas

often cannot provide sufficient means of survival, notably where rising population leads to reduced farm size. Poor rural families generally rely on a wide variety of on and off-farm activities and income sources. Many of these are based on natural resources. They include activities such as gathering firewood, preparing charcoal, fishing, hunting, handicrafts, and gathering non-timber forest products such as medicinal plants, fruits, and rubber, etc. Many landless poor also work as farm labourers. Survival and livelihood diversification strategies also include various types of migration. For example, some members of a household generally men may live semi-permanently in urban areas while others generally women usually stay in rural areas.

Addressing rural poverty and environmental degradation therefore often requires broad cross-sectoral approaches, which must go beyond agriculture. These efforts must focus on the diversity of livelihood sources and address systemic conditions that constrain the ability of rural poor to overcome poverty. For example, efforts to increase access to health, education and other basic social services, as well as transport and communication, directly affect the ability of rural poor women and men to pursue alternative incomes on and off the farm. Poor literacy and numeracy make it more difficult for rural poor to obtain information about ways to use resources sustainably and productively. It also limits their ability to develop livelihoods that do not depend on natural resources, or to obtain wage-earning jobs. Thus, improved access to education is critical to lower rural poverty and decrease the dependence of rural poor on natural resources for their livelihood. In many countries girl children are less likely to go to school, and more likely to drop out early, because of economic and cultural pressures. This differentially affects their livelihood opportunities, as they become adults. Similarly, access to effective reproductive health services is also needed to provide people with an ability to manage the size of their family. Issues related to health and education, although crucial, are not addressed in detail here, as the focus is on direct environment-poverty linkages (Figure 1). The interactions shown in Figure 1 are illustrative of certain poverty-environment processes where poor households are "compelled" to degrade environmental resources. However, this should not obscure the fact that much environmental degradation is caused by large-scale commercial operators, and by State policy; and there are many examples of positive actions by poor households and communities to manage environmental resources sustainably.

Nonetheless, there is a positive linkage between increased incomes and increased access to health and education, which in turn is likely to have positive impacts on livelihood diversification. Livelihood comprises the capabilities, assets (including both material and social resources) and activities required for a means of living. A livelihood can be said to be sustainable when it can cope with and recover from stress and shocks, and maintain or enhance its capabilities and assets, without undermining the underlying natural resource base. By placing people at the centre of development, a 'livelihoods' approach stresses the importance of influencing policies and institutional arrangements so they support the needs of the poor. For example, achieving conservation objectives in an area has often meant restriction or prohibition of traditional activities of importance for the poor. This raises two problems: Firstly, it is clear that the poor may be harmed when management demands less use in the short-term in return for a better resource later without access to alternative resources in the meantime. Secondly, such a policy may even be pointless from an environmental perspective for instance, if access to fuel wood from one area is restricted but no alternative provided to the poor, then this may simply shift the resource strain on neighbouring forests. Thus, the concept of "environmental entitlements" addresses the important issue of the extent to which households, especially those highly dependent on natural resources for their livelihoods, actually have adequate access to those resources. The sustainable livelihoods approach stresses participation by the poor in order to identify the key constraints they face and to seek the most promising alternatives.

Until the 1970s, poverty was usually understood in terms of falling below certain minimum levels of food intake, income or consumption. This narrow approach is no longer widely accepted, either by social scientists or by development agencies and practitioners. Progressively more significance has been given to the ways poor people themselves view their situation, and how poverty is conceived in different cultures. Today, a poor person is one who is unable to meet the minimum conditions of well being, as these are understood in societies around the world. Typically, this involves a range of inadequacies in consumption, several forms of insecurity and an inability to participate in what are considered minimal ways in social life. According to a recent view from the World Bank "poverty is multi-dimensional, extending from low levels of health and lack of education, to other 'non-material' dimensions of well-being, including gender gaps, insecurity, powerlessness and social exclusion". For the United Nations Development Programme (UNDP), human development is defined as "the process of enlarging human choices". Human poverty means therefore "that opportunities and choices most basic to human development are denied to lead a long, healthy, creative life and to enjoy a decent standard of living, freedom, dignity, self-respect and the respect of others". Given the multi-dimensional nature of poverty, the precise measurement of who is poor cannot only be restricted to measurement of income-based poverty, but must also define social categories such as gender, ethnicity, location, livelihood status, etc. A distinction can also be drawn between people in chronic (long-term) poverty and those suffering transitory poverty. The latter may be as a result of natural or macro-economic shocks. Households in chronic poverty can also be disaggregated according to their specific characteristics and causal factors. One category of the chronically poor may experience poverty in several dimensions, for example socially excluded groups, people with disabilities, refugees and displaced persons, people suffering from HIV/AIDs etc. Another categories are those who suffer chronic poverty primarily as a result of inadequate access to productive assets. This variation in the dimensions and characteristics of poverty highlights the need for careful analyses using different sources (including participatory poverty assessments) to support the development of appropriate interventions. Also, different manifestations of poverty have different implications for the environment. The table 1 below attempts to characterize poor households by rural/urban location and by the chronic or transitory nature of poverty. These categories are illustrative rather than giving a complete picture. Also, given the dynamic nature of poverty processes, there can be movement between the categories, e.g., a poor smallholder in one year may next year be part of the urban informal poor; a transitory poor formal sector worker may slide into chronic poverty through suffering long-term unemployment.

In India a large section of planners and policymakers think that there are no serious environmental problems among rural poor. According to them, if however there be any, a variety of solutions exist. They often are of the opinion that government should come forward to implement the solution. As a remedial package minimization of public sector interference and maximization of government intervention, ensuring suitable prices for infrastructure and urban amenities through removal or reduction of subsidies, generating employment directly through anti-poverty programs, easing private and joint sector projects are being prescribed. However the ultimate solution is being sought through competent and transparent management of environmental projects. Although the desirability of growth is globally accepted and recognized, recent years have provided testimony to raising anxiety about whether environmental constraints will limit development or whether development will cause serious environmental degradation. A number of environmental problems are already very serious and call for immediate attention. Environmental threats evolve because of both resource depletion and negative externalities caused by development process and projects. With a view to achieving the object of sustainable development, it is inevitable to do with both types of problem adequately (Sarkar, 2005).

PRESSURES ON ENVIRONMENTAL RESOURCES

Even for land-owning households, farming alone often cannot provide sufficient means of survival, notably where rising population leads to reduce farm size. Poor rural families generally rely on a wide variety of on and off-farm activities

Table 1

Characterization of Poor Households by Rural/Urban Location and by the Chronic or Transitory Nature of Poverty

Poverty "Type"	*Rural poor Urban poor*	*Rural poor Urban poor*
Transitory	Viable farmers, pastoralists and landless rural workers affected by sudden natural shocks or seasonal income losses; rural households affected by civil conflict	Urban formal workers (and dependents) suffering temporary unemployment or real wage declines
Chronic, asset-poor	Smallholders, small herders or landless with limited/ no access to productive assets, and/ or deteriorating asset base; rural communities isolated from markets and services; labour-poor households	Urban informal sector with limited/ no access to productive assets; labour-poor households, especially-female-headed households
Chronic, multidimensional	Marginalized cultural categories (ethnic or caste groups, marginalized indigenous communities); households with high dependency ratios; chronically sick and disabled	Marginalized cultural categories; households dependent on anti-social activities; refugees, internally displaced persons, undocumented aliens.

and income sources. Many of these are based on natural resources. They include activities such as gathering firewood, preparing charcoal, fishing, hunting, handicrafts, and gathering non-timber forest products such as medicinal plants, fruits, and rubber, etc. In sub-Saharan Africa, a 30-50 per cent range of reliance on non-farm income sources is common; in southern Africa, however, it may escalate to 80-90 per cent. In South Asia, on average, roughly 60 per cent of rural household income comes from non-farm sources. However, this proportion varies widely between, for example, landless households and those with access to land for farming (Ellis, 1999). Many landless poor also work as farm labourers. Survival and livelihood diversification strategies also include various types of migration. For example, some members of a household generally men may live semi-permanently in urban areas while others generally women usually stay in rural areas.

Table 2
Land Use Pattern in India

	1958-59		*1995-96*	
	Area ('000 hectares)	*Percentage*	*Area ('000 hectares)*	*1993-94 Percentage*
Total geographical area	2,93,972	100	3,04,863	100
Forest	52,675	17.9	68,421	22.4
Non-agricultural use	13,563	4.6	22,035	7.2
Cultivable waste	20,610	7	14,468	4.7
Permanent pastures and grazing	13,112	4.5	11,176	3.7
Other fallows	12,286	4.2	9,703	3.2
Agricultural use	1,43,136	48.7	1,56,428	51.3

Note: Agricultural uses = Current fallows + Net cultivated area
Source: Nadkani M. V., (1999), Environment in Karnataka: A Status Report, *Economic and Political Weekly*, XXXIV (9): 2735-2743.

Agriculture constitutes the major use of land in the country as a whole. Clubbing together net sown area and present fallow, it accounted for 48.7 per cent of total geographical area in 1958-59 and 51.3 per cent in 1995-96 in India. Though the proportion of land under agriculture increased by 2.6 per cent in India over this period, the proportion of land under forest with regard to legal status (not essentially under actual tree cover) enhanced in the country as a whole from 17.9 per cent to 22.4 per cent during the same period (Table 2).

Forest on the other hand as a natural resource is not only aggregate of all trees but also a whole of the living and non-living elements which has a cultural and spiritual value. According to a release issued by the Food and Agriculture Organization of United Nations (1995) global forest, woodland and scrub cover decreased by 2 per cent or 100 million hectares during 1980s. In developing countries though the forest cover decreased by only 98.7 million hectares in 1980s, the loss of natural forest cover alone was 163 million hectares i.e., 8 per cent during the same period. In India despite the initiation of massive plantation programs the actual forest cover is coming down over the years. It has been projected that annual logging of closed forest in India during the period 1981-1990 is 42,000 hectares i.e. 1 per cent of closed forest. The Indian Remote Sensing Satellite imagery (1995) reveals that the extent of forest cover of the country was over 19 pert cent of the geographical area, but the imagery also discloses that less than 12 per cent was dense cover. Therefore this deforestation has a lot of ill effects like loss of sustainable logging potential and of erosion protection, watershed stability and carbon sequestration provided by the forests. Further loss of bio-diversity tends to potential loss of modern drugs and genetic resources.

Substantive pastures, cultivable wastes and other fallows combined together can be said to constitute common property resources used for grazing. According to official data, the proportion of such land reduced significantly from 15.7 per cent to 11.6 per cent between 1958-59 and 1995-96 in India. The role of such common property resources as a source of fodder had been decreasing over the years. With the growing trend of population in urban areas and economic development price of livestock products had increased significantly over the years. So trend from official data shows that common property resources appear to have slowed down both in quantity and also in quality and productivity.

Poor households are often highly dependent on "common property resources" (CPRs), which include fallow fields, forests, fishing grounds, pastureland and wetlands for their livelihoods. CPRs are a source of a variety of goods including food, fodder, fuel, medicinal plants, which are important sources of sustenance or income for many land-less poor (See Box 1). For many rural poor women and men, CPRs are their main source of food, fuel, building materials, and income. For others, they are a critical source of supplementary income or food in times of crises such as drought, in periods when employment opportunities are scarce, or when food stocks are low before the harvest. Heavy reliance on CPRs thus makes poor women particularly vulnerable to their degradation, depletion, appropriation and/or conversion to other uses. Women, who often are not allowed to own land, make particular use of the resources found in CPRs for household purposes. Depletion of these resources means that women

Box 1

Common Property Resources (CPRs)

CPRs include uncultivable or fallow fields, pasturelands, forests, inland waterways, ponds, and low-lying wetlands. They represent a significant component of the land resource base and have special importance for the poor 1. The range of products drawn from CPRs for subsistence needs or for sale is wide and varied. It includes food, firewood, small timber, manure, fruits, medicinal herbs, roots, leaves, bark, fibres, seeds, nuts, gum, spices, resin, sap, syrup, oils, materials for house construction, handicrafts, etc. Hunting and trapping of mammals, aquatic species and birds often represent important food sources. Fodder and water for livestock are often drawn primarily from CPRs. Wetlands provide a special case in point. They are used by fisher folk, hunters, charcoal makers, pastoralists and agriculturalists under traditional resource sharing regimes at different seasons, and also harbour a wide variety of fauna and flora. Ensuring compatibility among such a wide set of users poses special challenges. Conventional access rights are particularly hard to define, since water levels are not the same each year and patterns of flooding are erratic. Even the most elaborate traditional arrangements seldom extend to upstream water users, often leading to over-extraction or pollution by industrial or agricultural users. This has severe consequences on downstream fisheries resources.

will need to walk further to collect water and fire wood, etc. Women and men tend to use different aspects of CPRs. Where women do not have the voice to directly influence decisions affecting the uses of CPRs, their exclusion from decision-making can result in resource uses, which negatively affect them.

The expansion of agriculture to open access forestlands is a major environmental consequence of poverty, food insecurity, and landlessness in many countries, notably in the tropics. Although nominally managed by the state, forests made accessible by the construction of roads become de facto "open access" in the wake of logging operations by private concessionaires, who often fail to comply with their obligations to manage the forest sustainably after harvest. Where there is a shortage of alternative income opportunities off-farm, landless peasants resort to converting these de facto open-access lands to subsistence agriculture.

Soil erosion is a principal problem affecting particularly in dry and rain fed areas. It reduces fertility of land owned by poor cultivators and invites drought. Apart from field productivity losses, soil erosion increases to costs such as river transport channels and other hydrologic investments. As far as soil conservation is concerned only 20.9 per cent out of total problem area of 52.2 per cent were taken into consideration up to 1992-93. So there is still a long path to go.

Besides water shortages add health hazards and it may be marked as ill effect of water degradation on productivity. It has already been estimated that by 2025 thirty-four countries including India will face water stress and of late about 29 countries are suffering from water scarcity. It is also expecting that number of people dwelling in water scare countries will increase from 132 million in 1990 to between 653 million (with a low population growth projection) and 904 (with a high population growth projection) in 2025. Water pollution on the other hand propagates many water borne diseases like diarrhea, hook worm, cholera, typhoid, paratyphoid etc. Moreover GREEN India Report of Tata Energy Research Institute (1997) reveals that from 1947 to

1997 availability of fresh water decreased by two thirds and area covered by soil erosion rose by in the neighborhood of 80 million hectares. On water pollution the report indicates that class one and class two cities generated around 20 billion litres of sewage wastewater per day but treated only a-tenth. The total sewage generation from urban areas increased six times in the last fifty years (1947-97). The water need of major water consuming industries like agro-based units, refineries, petrochemicals, fertilizers and chemicals had grown 40 times but these were not treating a large quantity of generated wastewater. Poor water management, combined with high population density compared to available water supplies, results in water shortages in many places. Shortages often have an impact on the poor first and foremost, as they may have intermittent irrigation water and be unable to grow crops reliably. The poor often have to spend many hours collecting water for domestic use, which can significantly limit their ability to generate income. This is a problem for women (and girl children) in particular, who are generally given the task. It can further limit the ability of female-headed households to cultivate land, since the women need to divide their limited time between collecting water and farming. Pollution from industry or urban centres is an important indirect cause of degradation of water bodies and lands. This increases the risk of exposure to toxic chemicals and disease pathogens either directly or through consumption of contaminated fish and shellfish. Women suffer the greatest exposure risk from polluted water because they contact water more than men do. Pollution also lowers the productivity of freshwater and coastal ecosystems, which directly affects the income and livelihoods of poor men and women using these resources.

Solid waste and hazardous wastes cause risks, which are often acute. Solid waste includes all types of solid and semi solid waste products like ashes, garbage, house sweepings etc. It enhances rat population and propagates many diseases. While dry waste is not disposed of local people burnt it in unscientific manners resulting in smoke and toxic gases in the air. Apart from that hospital waste also deteriorates health standards among city dwellers. Nursing homes, clinics, even bigger hospitals dump waste untreated. In lieu of acting as healing centre, hospitals themselves are becoming a source of disease. Moreover untreated wastes are being scattered by the birds and animals from the dumped sites, which often degrade environment. Diseases spread by rotting garbage and blocked drains are equally dangerous. Further these solid wastes create pollution of ground water resources. It has already been reported that municipal solid waste has increased seven times while the disposal of solid waste continues to be unscientific. It is estimated that generation of solid waste in major Indian cities is about 50 kgs per capita per day, while collection and disposal of such waste by the municipal corporation are almost absent.

Efficient municipal collection and disposal systems are essential components to urban environmental management, whether they are in the private or public sector. Developing formal waste management systems that build on the skills and knowledge of informal waste collectors can have important economic, social and environmental benefits. It can help increase the recovery and recycling of valuable resources found in urban waste streams, reduce residual waste volumes to be disposed of, and improve working conditions and incomes for informal waste pickers. This requires paying specific attention to the needs and constraints of informal male and female waste pickers who are often left out of decision-making processes. Instead, decisions concerning waste management generally focus on the priorities of middle and upper income groups who generate the largest volumes of wastes. However, increased use of informal waste pickers must at the same time reduce their exposure to the handling of hazardous wastes, for which specific urban environmental management plans and policy measures are required.

Many developing country towns model their waste management systems on those of developed countries, resulting in socio-economically inappropriate technology choices. Examples include waste collection vehicles that cannot cope with narrow or unpaved roads and lanes, or composting or incineration plants that are unsuited to the volume and composition of waste streams. Similarly, compacting waste reduces the possibilities to reclaim and recycle material, unless re-usable or recyclable material is specifically separated before waste is collected.

The livestock population, however, has continued to grow in the country though not keenly. Total livestock excluding poultry increased by 32.3 per cent in India between 1961-1990. The National Family Health Survey (1992-93) reveals that household having livestock in urban areas is 14.3 per cent. The density of livestock per hectare of land clubbing together forest, agricultural land including present fallows, pastures and cultivable waste was only 1.71 per cent in 1990 in India as a whole. The growth of poultry has been more significantly than of livestock. The total number of poultry increased by 141 per cent between 1961 and 1990 in the country as a whole. Livestock pressure on land, number of mammal and bird species and also higher plant species threatened having an adverse effect on regeneration capacity is a traditional source of worry to the environmentalists and foresters in India (Table 3).

LOSS OF BIODIVERSITY

Biodiversity loss is directly threatened by desertification, but is a much wider process observed in all major ecosystems. Biodiversity loss in areas experiencing desertification occurs as a result of the extensification process, whereby farmers attempt to compensate for declining productivity by converting more natural ecosystems to agricultural use, destroying habitats of animals and plants in the process. This is a critical issue for food supplies: almost all the globally important cereal grains originate from drylands, and the loss of the genetic forebears of these food plants could impair future ability to adapt their genome to accommodate a changing environment.

Biodiversity values vary significantly in type and amongst different stakeholders. Box 2 shows the different types of use and non-use values of biodiversity. Use values are of most immediate interest to poor households, in terms of supply of consumption items and income generated from marketable crops and animals; but the ecological services provided by biodiversity, and the risk spreading effects of crop diversity, are also valued by poor households.

Box 2
Biodiversity Values

Direct use:
- Commercial goods – natural ecosystem products, agricultural products, non-consumptive use.
- Subsistence and barter goods – food, fuel wood, building materials, medicines.

Indirect use:
- Ecological services – soil retention, water filtration, air cleaning, carbon sink.

Option values:
- Risk reduction through crop diversification
- Potential value as a resource that might yet be unknown

Socio-cultural values:
- Self-sufficiency/autonomy for people
- Integral part of cultural identity and common heritage; symbolic/ aesthetic; religious

Intrinsic/Bequest value:
- Maintenance of future options

Some of the benefits of biodiversity, in contrast, appear to accrue to other stakeholders, including biotechnology companies, as well as the global community, which gains from the absorption of atmospheric carbon in well-forested areas. There may be potential conflicts between land users who prioritize increasing productivity through reducing plant diversity, and the broader global community. The Convention on Biodiversity recognizes these tradeoffs and emphasizes the need to maximize the social and economic benefits from the protection and sustainable use of biodiversity, and their equitable distribution. There is some evidence that agricultural biodiversity is greater in areas, which economically are regarded as poorer; but this does not imply a causal relationship. Biodiversity in such areas is particularly important for poorer households for spreading risks, enabling such households to generate livelihoods from a range of products. For marginalized groups maintenance of, and improved access to, agricultural biodiversity can contribute more to sustainable livelihoods than can conversion to cropping patterns with reduced diversity in part because these groups' traditional entitlements to such biodiversity may be stronger than their market access to the production inputs needed to support more "intensive" agricultural systems. One of the key requirements to maintain biodiversity is to reform the system of economic incentives, institutional and policy structures, which are currently geared in favour of, industrial-type agricultural models and against systems promoting agricultural biodiversity. Current incentive systems, for example, provide distorted signals by failing to reflect the external effects of biodiversity loss.

AIR POLLUTION

Air pollution has many perilous chronic hygienic impacts. The worst pollutants of air include carbon monoxide, sulphur dioxide, nitrogen oxide etc. Automobiles, industries, electric power plants and households are the main sources of air pollution. Indoor and outdoor air pollution breaks out various air borne diseases like pneumonia, bronchitis, asthma and even lung cancer along with many other problems. Vehicle emissions have been identified as the greatest environmental danger in a large number of cities in developing countries. So far as carbon dioxide emissions are concerned, in Table 4 country wise percentage of global carbondioxide emissions reveal that a list of 12 countries taken together contribute 62.8 per cent of total carbondioxide emissions, out of which USA alone share 23.4 per cent of world emissions. Hence, almost one-fourth of the global carbondioxide emissions come from USA alone. Besides, China and India contribute 14.9 per cent and 4.4 per cent of global carbondioxide emissions because of their high population, while their per capita emissions are only 2.8 and 1.1 metric tons respectively compared to 7 to 20 metric tons per capita emissions for rich income countries. Although the low-income countries contribute a high percentage of emission, but as far as per capita emissions are concerned, they contribute a low level as compared to rich income countries. It is interest to note that there is a link between carbondioxide emissions and average annual growth rate of GNP. Countries like China and India with a high average annual growth rate of GNP have largest share of emissions. Indonesia contributes a little percentage of emissions of 1.1 having average annual growth rate of (-) 14.8 per cent. It has further been observed that there is a far correspondence between high percentage of world GNP and percentage of world carbondioxide emission. By and large, countries having a large share in world GNP contribute larger percentage in emission. Obviously, mighty countries are definitely responsible for the emissions of carbondioxide in the atmosphere. Studies run by National Environmental Engineering Research Institute reveals that the level of pollution due to round the clock automobile emissions in Indian cities is on the rise if compared to other cities of the

Table 4
Country-Wise Share in Global Carbondioxide Emissions and Average Annual Growth of GNP

Country	*Carbondioxide emission (% of world)*	*Carbondioxide emission (per capita metric tons)*	*Share in world population (%)*	*Average annual growth rate of GNP*	*Share in world GNP (%)*
USA	23.4	20	4.6	3.7	27.4
China	14.9	2.8	21	7.4	3.2
Japan	5.2	9.3	2.1	(-) 2.6	14.1
India	4.4	1.1	16.6	6.1	1.4
Germany	3.8	10.5	1.3	(-) 0.4	7.3
UK	2.5	9.5	1.0	2.0	4.3
Italy	1.8	13.7	0.5	6.1	0.4
Canada	1.8	7.0	0.9	2.3	4.0
Australia	1.4	16.7	0.3	3.8	1.1
South Africa	1.3	7.3	0.7	0.6	0.4
Brazil	1.2	1.7	2.8	0	2.6
Indonesia	1.1	1.2	3.4	(-) 14.8	0.4

Source: World Bank (1999), *Entering 21st Century,* World Development Report 1999/2000, Washington DC.

world. Particulate emitted by vehicles pose a hazard to health of human beings, animals and also to longevity of the property. However the damage due to particulate is rather indirect and slow but among the gaseous components, oxides of nitrogen and nitrated organic are considered to be most hazardous. Even very small quantities of these chemicals cause problems like irritation of eyes, nose and other delicate membranes of the body. While not so dreadful, carbon monoxide is also hazardous to human beings and animals. It reacts and neutralizes of a portion of the hemoglobin in the blood and thus reducing respiratory capacity.

Table 5

Adverse Impacts of Industrial Pollution on Human Health

Industry	*Pollutants*	*Damaging impacts on human health*
Cement	Cement dust	Asthma and other bronical problems
Leather	Sulphur oxide and other acid gas	Suffocation, bronical problems
Paints	Sulphur dioxide, hydrogen sulphide	Suffocation, irritation of throat and eyes, irritation of lung, blockage of oxygen and sticky suspended particulate matter
Aluminum	Ash, acid waste and sewage water	Blood poisoning, suffocation, lung cancer, health diseases
Iron and steel	Slag, dust and sludge	Headache, corrosion of teeth, cold, eye defects, lung diseases
Fertilizer	Urea dust, ammonia, phosphate, solid waste-fly ash	Suffocation, bronchitis, edema of lungs, irritation of throat

Source: Choudhury, A and Sahu, N.C. (1999), Environmental Excise Duty and Emission Tax: A Comparative Analysis in the Context of India, *Finance India*, XII (1): 117-134.

The pollutants emitted by selective industries like cement, leather, paints, aluminium, fertilizers, sugar and paper have damaging effects on human health, cattle health, forests and biodiversity. The diseases namely asthma and chronic bronchitis, respiratory troubles, teeth and gum problems, eye and ear diseases and so on frequently attacking the people of industrial regions are attributable to the air pollution caused by industries. The health damage due to environmental pollution has been identified in many studies. Meanwhile all most all state pollution control broads have found pollutants of selective industries and their adverse impacts on human health (Table 5).

Smoke and fumes resulting from indoor use of biofuels jeopardize health standards more than any other outdoor degradation. In India the most important disease associated with indoor and outdoor air pollution is probably acute respiratory infection. The acute respiratory infection among children below five years of age is 86 per cent in India and at global level it is also single largest disease category. A balance sheet of human development (1990-97) highlights that every year nearly 3 million people in the world fall prey to air pollution, out of which more than 80 per cent from indoor air pollution. An analysis of National family Health Survey data reveals that adults over 30 years living in households using biofuels had around 30 per cent more partial blindness and 170 per cent higher tuberculosis rate than those living in households using clean fuel. Estimates available from a recent study on annual premature death to children under 5 years of age and adult women are in the range of 4,10,000-7,90,000 (Table 6). These estimates are however only for specific diseases there are certainly on other population groups and from other disease also.

Table 6

Estimated Annual Premature Deaths from Indoor Air Pollution

A. Strong evidence	3,10,000-4,70,000
Acute respiratory infection	Children age below 5 years
Chronic obstructive pulmonary disease	Women
Lung cancer from coal use	Women (few in India)
B. Moderate evidence	50,000-1,30,000
Blindness	Women (no death)
Perinatal effects	Insufficient data for estimates
Tuberculosis	Women
C. Suggestive evidence	50,000-1,90,000
Cardiovascular disease	Women
Asthma	Women (few in India)
Gross total of all three categories	4,10,000-7,90,000

Source: Smith K. R., (1998), *Indoor Air Pollution in India: National Health Impacts and Cost Effectiveness of Intervention,* Report Prepared for Capacity 21 Project of India.

DATA AND METHODOLOGY

In the present endeavour six villages of Jalpaiguri district of West Bengal, India have been selected using multistage sampling design to study environmental impact of pollution on rural poor women in general and health related diseases in particular and to assess the data in comparison with the earlier survey conducted in 2004. The villages were selected to collect detailed information regarding the losses and damages due to environmental pollution. The information was collected through a structured questionnaire and informal group discussion (participatory rural appraisal method) with the rural communities in the villages. A sample of 50 women respondents was drawn using stratified random sampling techniques. This was the second phase household survey being conducted during May-June 2007 followed by first phase during May-June 2004. The survey was carried out at three levels, viz. household level survey including individual responses for health status, village level and health centre level survey. At the household level multi pronged approach was used to gather different types of information relating to perceive human health. Village level information was collected from the head of the village *panchayat* whereas health related information was collected from administrative office of the concerned health care centre.

FINDINGS

The study of selected villages reveals that women and children are badly affected by pollution. It is found that most of the diseases are indoor air and water borne, such as bronchitis, tuberculosis, asthma, eye disease, adverse pregnancy outcomes, skin infection, joint pain, other respiratory diseases, diarrhea, malaria etc. Also general muscular weakness, phlegm and cough among children have been noticed in the villages. A majority of complaints are about problems regarding bronchitis and tuberculosis. It is also reported during the period of survey that eleven women admitted to the hospital due to drinking contaminated water. The women suffering from respiratory diseases and water related diseases appear to be very high. The women suffering from respiratory diseases are 12 per cent (11 per cent in 2004) and from water related diseases are 9 per cent against 8 per cent in 2004. Asthma was reported by 4 percent (2 per cent in 2004)). The reporting of symptoms and diseases were kept into account on the basis of respondent's memory recall for all the members of the households. Incidence of asthmatic symptoms noticed to be higher among women who cook inside compared to those who cook open air. An adjusted odds ratio of 2 for blindness among women was found in biomass using households. Adverse pregnancy outcomes (still birth, low birth weight) and early infant death have been linked with outdoor air pollution and passive smoking, and it is found that odds ratio in this case has been increasing (2.67 in 2004). On the basis of self reported symptomatic cases of respiratory diseases cough, breathlessness, wheezing are found to be significantly higher among those living *kachcha* houses and using biofuels for cooking compared to those living *pucca* houses and using clean fuels for cooking as found in 2004 survey. The records from health centre disclose that five health centres received more than 30 (two health centres less than 20 in 2004) female patients in a day. Average number of female patients visited in a day by all the surveyed health centres together was 43 against 37 in 2004 survey and average number of female patients treated in a month by these health centres for respiratory or water related problems was 294. The survey signifies that percentage of affected women in each household is higher because women do with entire household work with contaminated water. The average number of days sick and unable to work every woman is about 30 (38 in 2004) a year. It is found that 5 per cent against 3 per cent in 2004 of sample women are suffering from worm infection. Many women were found who have been victimized due to arsenic poisoning. Besides diarrhea and malaria are common among a few women and children in the survey villages and increased over last three years.

CONCLUSION

The livelihood options available to poor women and men in rural areas include, but are not restricted to, the use of natural resources. Poor households engage in a wide range of other small-scale or micro-enterprises, for example to meet local demand for semi-processed food products, household goods, farm implements, and services. For most households these activities are additional to on-farm production but may in fact be the main source of cash income. As pressure on natural resources increases, the significance of off-farm activities also increases, and expanding the scope for such activities will help to relieve resource pressures. Policy measures therefore need to address factors, which may restrict the growth of rural enterprises, including lack of roads and other basic infrastructure, limited access to credit, etc. Improving education and physiçal infrastructure may have greater impact on expanding income opportunities for the landless poor than investments aimed at enhancing on-farm production. Nonetheless neither should be supported at the expense of the other since both are clearly required. Any measures aimed at stimulating development of off-farm enterprises should pay specific attention to the needs of poor women, who in many areas have shown their ability to develop successful micro-enterprises whilst maintaining loan repayment rates higher than those of men.

Attacking poverty while enhancing the environment requires first and foremost the political will to eradicate poverty. In most cases, this will ultimately require some redistribution of resources or rights of access to resources - toward poorer sectors of society. The search for win-win situations should not divert the state from also reallocating resources towards the poor. The political will of the state in championing the cause of the poor is tested through its commitment to uphold and enhance rights regimes for the poor. Community-based decision-making and transparent dialogue cannot occur without political will at the highest level. While all rights regimes involve some regulatory and supervisory role for the state, in some cases the rights regime may already be relatively favourable for management by the poor, but they are not able to protect their rights. Poor people need to be protected from or empowered to fight against actors who encroach upon their resources. For example, coastal fisher-folk may have the right to fish in certain waters, but they may be powerless against trawlers that sweep through their fishing grounds. People dependent on the forest may have no recourse against logging taking place in upper watersheds, yet they bear the consequences in the form of floods, drought, and soil erosion. Protecting the asset base requires a wide range of actions, including support to strengthening community-based organizations, which genuinely represent poor people, and actions in the regulatory, and governance areas, which strengthen the legal basis of poor people's rights. Civil society organizations can often play a role in supporting communities in asserting their rights over resources. Beyond protection of resources, there is a need for support to develop the asset base of the poor, including the environmental resources which poor people are heavily dependent on. The protection of resources may not be enough to ensure that resource users can enhance their livelihood options, especially where population growth is increasing pressure on the resources. There is a need for technical and institutional innovations, which support communities in augmenting the resource base (e.g. grazing land enrichment, fisheries management measures which enhance fish stocks, joint forest management, etc.), and/or in finding substitutes, which reduce the pressure on resources (e.g. alternative fuel sources, expansion of off-farm income-earning opportunities etc.).

In India several regulatory measures have been initiated regarding pollution control and environmental encouragement. The rapid changing scenario exhibits that India has learned to depend more on markets and less on government to accelerate development process, the result of which has been most unfavourable. However in order to adding momentum to the development process and to implement strong environmental policies the intervention of government becomes indispensable in the present situation. Usually two sets of policies are advocated to fall on the underlying causes of environmental degradation and both are equally essential. The first set of policy packages includes promoting education, family planning and poverty removal programs, facilitating environmental management research etc. The purpose of second set of policy packages is to fetch the positive links between development and environment. The most common among them include targeted policies to change behaviour. In second set policy packages are based on both incentives and quantitative restrictions. However the government adopts various fiscal measures taking into account incentives based policies to control degradation. Among them, use of environmental taxes is regarded as a part of an integrated policy that has become a subject of considerable interest in most of the countries. The basic facility of environmental taxes is that it can prevent and control degradation to the possible extent on the one hand and ensure availability of funds for environmental programs on the other. The government should impose environmental taxes on those inputs, which result in pollution. Apart from the environmental taxes some countries are moving towards carbon tax to fight with the green house gases or global warming. The imposition of carbon tax helps the customers to use less energy and switch over to less carbon intensive energy sources. However the introduction of environmental taxes and integration of environmental charges with the tax policy requires a set of simple rules and criteria to ease their implementation and strengthen their effectiveness. The government often takes measures to encourage economic efficiency on both national and international levels to minimize the degree of environmental pollution, but the measures so taken are not equal to the need. The National environment policy is designed in such a manner that those policies can provide the benefits to the individual countries only, not the rest of the world. As a result the global environmental goals are not achieved at par. Therefore government must consider the implications of indigenous environmental policy decisions for other countries for successful international cooperation in environmental arena. Side by side apart from the government machinery to enforce it is inevitable to take care of growing environmental threats in regular, rational and systematic ways and consciousness to environmental concerns is to be built up among the general masses by the government (Sarkar, 2005).

Poor people are often constrained in their ability to influence government decisions that affect the resources on which they depend. This creates uncertainty about their future access to resources and undermines incentives to use their resources sustainably. Specific efforts are required to enhance their full participation in decision-making processes that will affect their resources, e.g., citizen oversight boards and community-level review processes for dispute resolution. Governments often neglect the gender dimensions of decisions made on resource use, and particular efforts need to be made to enhance women's participation in decision-making processes. In many cases, the poor already have the formal right to manage key environmental resources, but they are not able to protect their rights, either because they are not aware of them or because they are unable to defend them. Individuals and communities need to be informed of their rights and of the scope for influencing government policies through participation in political processes. Government institutions must also be proactive in disseminating information, seeking public input, and in working with local communities to identify and meet local needs. Many communities have become more effective in interacting with government decision makers after being trained in such skills as talking in public fora, negotiating, and organizing public meetings. Civil society organizations can often play a critical role in supporting community-level capacity development for protection and sustainable management of natural resources. Women often face special difficulties participating in local political processes. Social constraints often prevent them from attending and/or speaking openly in meetings with men, or men ignore the women when they do. Efforts to involve the rural poor in governance will need to explicitly involve women. Government officials often need to better understand the social and ecological constraints faced at the local level in order to support local resource management. Capacity development efforts can assist this better understanding, which should increase political will to avoid policies that lead to misuse of resources or inhibit the rural poor from improving their livelihood. Co-management of natural resources between the State and local resource users provides a framework for sustainable management, which at the same time enhances the livelihoods of poor people.

REFERENCES

Angelsen A. (1997), The Poverty-Environment Thesis: Was Brundtland Wrong? *Forum for Development Studies*, (1): 135-154.

Choudhury A., and Sahu N. C. (1999), Environmental Excise Duty and Emission Tax: A Comparative Analysis in the Context of India, *Finance India*, XII(1): 117-134.

Ellis (1999), *Rural Livelihood Diversity in Developing Countries: Evidence and Policy Implications*, Overseas Development Institute, London.

Nadkani M. V. (1999), Environment in Karnataka: A Status Report, *Economic and Political Weekly*, XXXIV (9): 2735-2743.

Sarkar S. (2005), Natural Resource Environment, Pollution and Women Health, In: T. Sabanna (Ed.), *WTO and the Agriculture*, Serials Publications, New Delhi.

Smith K. R. (1998), *Indoor Air Pollution in India: National Health Impacts and Cost Effectiveness of Intervention*, Report Prepared for Capacity 21 Project of India.

United Nations Secretariat on the Convention to Combat Desertification (UNCCD) (Undated), *Combating Desertification in Africa: Fact Sheet 11*, Bonn.

World Bank (1999), *Entering 21st Century*, World Development Report 1999/2000, Washington DC.

17

WOMEN AS ENERGY ENTREPRENEURS

Siddhartha Sarkar

INTRODUCTION

Access to better energy services can improve women's social, economic and political status - reducing the time and effort involved in household chores, providing better health and educational conditions, expanding income-generating opportunities, and easing their participation in public affairs. At the same time, greater sensitivity to gender issues increases the effectiveness of energy programmes and policies, as well as other types of development activities that involve energy use, by ensuring that the needs and concerns of both men and women are taken into account. Availability of kerosene or liquefied petroleum gas, improved stoves, electricity, and mechanical power significantly improves the quality of life for women in rural areas and relieves them of much of the difficult, unpaid work currently required to care for their families.

Decision-makers often view their energy related choices as gender neutral, however men and women are affected differently by energy policies wherever their home, work and community roles differ. Economic development policies thought to be gender blind might actually reinforce gender inequalities. For example, small amounts of electricity at home in the evening hours may improve the quality of life for some members of the family, allowing them to read or watch television, while for women, it may simply extend the burdens of the working day into the evening hours when they would otherwise be able to rest. Attention to these sorts of differing interests is needed in order to achieve effective and equitable distribution of energy and development services. Using a 'gender mainstreaming' approach ensures that the different impacts of an action or policy for men and women are evaluated so that gender inequality is not perpetuated.

THE PROBLEM

Gender sensitivity in energy policies is not just a matter of taking women's feelings into account. It involves acceptance of existing economic realities for women, and new financial investments in energy infrastructure and services that will best promote economic and social development for both men and women. However, women's roles as agents of economic development are often overlooked because so much of their labour is either unpaid or involves informal sector activities ignored in national economic statistics. Moreover, social roles, cultural restrictions, household burdens and child-rearing responsibilities tend to keep women at home and limit their options for outside employment. Women in rural areas of developing countries face particular challenges when their traditional roles make them responsible for growing food to feed their families and supplying household needs. Their ability to expand farming beyond a subsistence level or develop other kinds of businesses is often hindered because women generally do not have the same access as men to training, services, equipment, and financing, and in some cases men take over control of enterprises when they become profitable. Women are also often constrained by laws or social discrimination that prevent them from owning property, borrowing money, and gaining access to land and natural resources.

Despite these obstacles, many women are involved in income-generating activities, particularly small-scale home-based businesses compatible with their household and care-taking responsibilities. However, it is extremely difficult to sustain or expand these activities without electricity, running water, modern fuels for cooking, heating and lighting. Successful agriculture-based operations also require pumps for irrigation and mechanical power for processing harvests.

Other typical income-generating activities - processing food, baking, brewing beer, making soap and shea butter, and producing pottery-using kilns - require large amounts of thermal energy. Without access to modern fuels, women must collect large amounts of wood, dung, agricultural wastes or other traditional fuels. This detracts from activities that would directly improve their families' economic prospects.

Throughout the world, traditional divisions of labour keep many women and girls at home with responsibilities for childrearing and everyday household tasks. This makes it harder for them to participate in educational opportunities, outside employment, and community affairs.

In developing countries especially in poor, rural areas much of women's time is taken up with growing and processing food for their families without the benefit of any mechanical or motorized equipment. Women often cook over open fires using wood, dung and agricultural wastes they gather themselves and carry home on their heads, shoulders or backs (Sarkar 2006a). Where there is no pumped water, women and girls also go to wells, springs, lakes and rivers to carry back water for drinking, washing, cooking and sanitation. The drudgery and backbreaking work involved in these daily activities sap women's health and strength. In addition, women are exposed to a variety of health hazards from cooking over poorly ventilated indoor fires, including respiratory infections, cancers and eye diseases.

Girls may be kept home from school to help their overburdened mothers, thereby perpetuating conditions that reinforce female illiteracy and poverty. In urban areas as well, especially in 'informal' settlements, women are more seriously affected by a lack of basic services and infrastructure than men due to gender-based divisions of labour. The time and effort spent by women and girls on routine tasks could be reduced significantly if adequate resources were in place to deliver electricity, modern cooking and heating fuels, running water and sanitation, and basic transportation services. Reducing women's household burdens can have dramatic effects on women's levels of empowerment, education, literacy, nutrition, health, economic opportunities and involvement in social and political activities. Improvements in women's lives can bring about significant beneficial consequences for entire families and villages.

Energy planning that is intended to be implemented in a gender-neutral way can miss important issues and be inadvertently discriminatory against women. For example, a policy to promote the use of electricity by small enterprises may neglect the fact that many of women's income-generating activities (such as food preparation and processing, beer brewing and pottery making) require process heat for which electricity may not be the best option. In that situation, policies to promote effective distribution networks for liquefied petroleum gas (LPG) or other fuels would better support women's business needs.

For the most part, policy-makers do not take into account the different needs of men and women with regard to energy services. As a result, policies generally do not recognize that there is a gender bias, and women's energy needs tend to be marginalized in policy documents.

In areas without adequate energy services, women's priorities generally include lighting (in homes, for better working conditions in the evening and in the streets for public safety), cleaner fuels for cooking and heating, mechanical power and electricity for grinding grain, pumping water and supporting small business enterprises, and better transportation options. Women also benefit from policies that: promote better access to financing facilities and allow them to invest in equipment and pay for energy supplies; expand training programmes and extension services to improve commercial operations; and remove legal barriers that restrict women's property rights and limit their full participation in social and economic development.

Throughout the world, governments are moving away from publicly controlled energy sectors in favour of greater liberalization and competition. Although market-based energy investments provide important benefits, the private sector alone is not likely to meet the energy needs of rural areas in developing countries. Innovative approaches, designed using input from both women and men, are needed to remove regulatory barriers and provide incentives for communities to create their own energy solutions.

In what ways are women in developing countries particularly affected by lack of energy services?

- They grow food, maintain households, raise children, and operate small-scale enterprises without mechanical or electrical equipment or modern fuels.
- They have to find, collect and carry water and traditional fuels such as wood, charcoal, and agricultural wastes.
- They suffer health problems from hauling heavy loads, working over smoky fires, and giving birth without adequate health care facilities.
- They lose out on education and employment opportunities due to their household burdens.

ENERGY SERVICES AND MILLENNIUM DEVELOPMENT GOALS

Recognizing the importance of women's full participation in Development initiatives, the Millennium Development Goals (MDGs) adopted by the UN General Assembly in 2000 set international targets for promoting gender equality and empowering women. In addition, empowerment of women is a crosscutting issue that is critical for poverty alleviation and achievement of the other Millennium Development Goals.

Access to energy, too, is a critical factor for achieving the development goals relating to poverty alleviation, improvements in health and education, and environmental sustainability as well as the empowerment of women. Energy services such as lighting, cooking, heating and cooling, water pumping, refrigeration, transportation and communications are essential for economic growth and social advancement, and increased access to these services can help women move towards greater equality and empowerment.

In many places, traditional gender roles, financial subordination, lack of education, and social constraints limit women's ability to improve their economic and political status. Typical divisions of labour keep women in the home with responsibilities for everyday household maintenance tasks and child rearing, restricted in their ability to earn their own money and engage in civic activities. When women do engage in farming or business, they are often constrained by laws or social discrimination that prevent them from owning land, borrowing money, gaining access to natural resources, or receiving training and extension services.

Because of these socially determined roles and constraints, lack of energy has a disproportionate impact on women and girls who are living in conditions of poverty in rural areas, or in 'informal' urban settlements. Although access to more modern energy alternatives will not necessarily lead to greater equality in gender roles, it can at least relieve some of the most burdensome and unhealthy aspects of their daily lives and expand the development options available to women, their families and their communities.

Lowering exposure to indoor air pollution is critical for widespread advancements to the health of women and children in developing countries (MDGs 4 and 5). Women who are healthier will be better able to participate in economic enterprises, thereby reducing poverty (MDG 1) and increasing women's empowerment (MDG 3). A shift away from traditional fuels for cooking and heating practices will also free up time now spent collecting firewood, dung and crop residues, allowing more girls to attend school (MDG 2), and will reduce pressures on forest resources and allow crop residues to be used to fertilise fields (MDG 7). Since women and their dependent children make up the majority of those living in poverty, energy initiatives that increase income-generating opportunities for women can be important factors in reducing hunger and poverty levels (MDG 1), and can also enhance women's social and political status, and promote the empowerment of women (MDG 3). When women have access to adequate fuel and water supplies and money for school fees, their children are more likely to attend school, especially the girls since they will not have to be kept home to help their overburdened mothers (MDG 2). Money for food, and adequate fuel and equipment for processing and cooking it, also promotes better health for women and their families (MDGs 4, 5 and 6) and may reduce pressures on forests and other ecosystems vulnerable to overuse (MDG 7).

Since women are the primary food producers in many of the poorest regions of the world, increasing their access to motorised equipment and modern fuels for growing, processing, cooking and marketing food will help reduce overall hunger and poverty levels (MDG 1). Limiting the time and effort spent by women and girls in collecting fuel and water will allow them to take greater advantage of opportunities for education, employment, and involvement in political affairs (MDGs 2 and 3). Improving access to clean fuels for cooking and heating will lessen exposure to harmful indoor air pollution and ease the hardships of fuel collection (MDGs 4 and 5), as well as reduce pressures on fields and forests as sources of fuel (MDG 7).

MDG 1: Eradicate Extreme Poverty and Hunger

Since more women than men are living in poverty, ensuring that women's energy needs are met will reduce women's burdens in providing food and water for their families, and improve their ability to earn income.

MDG 2: Achieve Universal Primary Education

Girls are more likely than boys to be kept home from school because they are needed to help their mothers with the burdens of gathering fuel and water and performing domestic and agricultural chores, or because there is not enough money to send them.

MDG 3: Promote Gender Equality and Empower Women

Better access to energy services frees women from extreme household drudgery, increases their educational and employment possibilities, and allows them to participate more fully in community and political affairs.

MDGs 4, 5 and 6: Reduce Child and Maternal Mortality and Combat Infectious Diseases

Cleaner cooking fuels and improved stoves reduce respiratory illnesses that affect women and children. Pumped water from clean sources, and/or energy for purifying water, reduces the spread of water-borne diseases associated with child mortality. Energy services allow health clinics to have lights, water pumps, refrigeration for drugs and vaccines, medical instruments, fans and sterilisers.

MDG 7: Ensure Environmental Sustainability

Women, especially in rural areas, affect the sustainability of land and natural resources in their traditional roles as managers of wood and biomass fuel supplies and related environmental resources.

Although there have been energy-related initiatives targeted towards women, most have involved cooking and other household issues. Women also need energy for productive purposes to raise family incomes. Increased access to modern fuels, such as liquefied petroleum gas (LPG), kerosene and biofuels, can not only ease the time burdens and health problems associated with use of traditional fuels, but also enable women to initiate or expand small business enterprises. In rural areas without access to grid-based electricity–the places where women's opportunities are most limited–decentralized rural electricity systems are very helpful in expanding available economic options. Decentralized power can be obtained from conventional energy sources, such as diesel generators, or from village or household systems using renewable energy sources such as solar, wind, micro-hydro and modernized biomass technologies. When electricity is used to support small home-based enterprises, the extra income earned can make the service affordable for household as well as business uses, thereby improving the family's quality of life and their prospects for escaping poverty. In some cases, women can earn income by becoming energy entrepreneurs themselves – producing and marketing cleaner-burning stoves, renewable energy equipment, or alternative fuel products, such as biomass briquettes. Although increased income alone will not eliminate pervasive social and political discrimination against women, the self-reliance, expertise and confidence women gain from managing business enterprises can help them gain greater control of their lives in other ways as well.

- More efficient stoves to reduce fuel collection requirements.
- Access to modern fuels for cooking, heating and lighting, including liquefied petroleum gas (LPG); biogas and other biofuels, such as ethanol and biodiesel; biomass briquettes and, kerosene.
- Pumped water, provided by means of mechanical pumps (wind or water mills) or motorized pumps powered by grid electricity or decentralized generators run on biofuel, water or wind turbines, solar photovoltaic panels or diesel fuel.
- Equipment for agricultural production, grinding grain and processing food, sewing, refrigeration, lighting, communications and other household purposes, using modern fuels and motors, or electricity either from the grid or produced by biogas generators, hydro power, wind turbines, solar photovoltaic panels or diesel motors.
- Access to vehicles for going to markets and transporting goods, supplies and natural resources.

ENERGY AS WOMEN BUSINESS

Recognition of women's energy needs for household responsibilities and small business activities could make a significant contribution to developing countries' efforts to reduce poverty, raise health and education levels, and prevent pollution and environmental degradation. Furthermore, increased access to energy for women can create new economic and social opportunities for families and communities, while decentralised energy planning and delivery mechanisms can help empower women and strengthen their involvement in energy-related decision-making.

Mainstreaming Gender and Energy in Asia

A progressive change in the paradigm of development has taken place in Asia regarding the synergy between development goals, gender and energy. In practical terms, for policy makers and activists the integration of 'gender' and 'energy' has been difficult. Until recently energy has been considered primarily from the technical and commercial points of views, without much attention to its social context. Gender, however, is seen as a social construct that has evolved throughout human civilization, with features deeply rooted in cultures and traditions. In Asian societies there is great resistance to transforming the cultures under which men's and women's roles in a given society, environment and livelihood system have been established.

Most recently, in 2000 the UN General Assembly endorsed the Millennium Development Goals, which included the goal of promoting gender equality and empowerment of women as the third goal out of eight. In Asia, the strategic goals and global declarations that have been discussed in many forums have enabled countries and cultures to examine the gaps between men and women, and have provided windows of opportunity for integrating gender into mainstream development.

More than 50% of the world population of 6.1 million lives in the Asia Pacific region. Although total energy use in the region is high, per capita energy use is low and only marginally above sub-Saharan Africa. The total energy picture is important from two perspectives. The first involves how far the region can proceed in providing energy in order to further the implementation of the MDGs. The second is to what extent energy solutions in this region can provide feasible solutions to gender issues while dealing with the emerging concerns over development?

World comparisons reveal that in 2001 the industrial countries used 4.7 tonnes of oil equivalent (toe) per capita, while sub-Saharan Africa and the Asia Pacific region used only 0.8 toe per capita, less than 50% of the world average of 1.7 toe per capita. In a development context, and also from the perspective of eradicating imbalances in global equality, Asia should draw special attention. Meeting increasing energy demand, enhancing energy services for those who do not have access, and providing energy supplies for economic growth and poverty alleviation, are crucial issues. A breakdown of primary energy use by the regions reveals that in 2001, the Asia Pacific region used around 2.31 Gtoe (UNDP, 2004). In terms of primary energy mix, nearly 39% was derived from coal, 24.5% from crude oil, 24% from combustible renewable and waste, 7.3% from gas, 1.7% from hydro, 0.9% from petroleum products, 0.8% from nuclear and 0.5% from geothermal, solar and other renewable technologies. In regard to electricity generation, the general overview shows that out of the total of 2,795 Tera Watt Hours (TWh) of electricity 63.6% was generated from coal, 16.1% from hydro, 10.2% from gas, 6.6% from geothermal, solar etc. and 0.2% from combustible renewable and waste. Extended analysis reveals that energy usage varies greatly between the countries in the region. A substantial percentage of the total energy used in Asia is derived from biomass energy, providing heat and light, and in some areas motive power as well. In Asia, as in Africa and Latin America, the percentage of biomass in the total energy picture is higher than the worldwide percentage, which is around 11%. Biomass energy accounted for nearly 25% of the total primary energy of 93.7 exajoules

(EJ) a year. In terms of total final energy consumption it accounted for around 34.6% of 66.7 exajoules a year (ESMAP 2005). It is also important to note that biomass energy is primarily produced using traditional technology and unrefined solid fuels such as firewood, crop residues, animal dung and charcoal. Improved biomass energy technology or modern equipment is only used in a few countries.

Looking at the energy sector profile in Asia there is clearly a need for greater access to clean and affordable energy services by the poor, and in that context gender and energy linkages could be strengthened. Poor people and women are on the periphery of the energy system in many countries, particularly regarding per capita use, modern energy access and productive use of energy. There are strong correlations between low electricity consumption per capita and poverty concentration patterns. The electricity consumption per capita in developed countries such as Japan and Singapore is well over 100 times higher than in the countries like India, Pakistan, Sri Lanka and Vietnam where GDP per capita is low. The Asia region accounts for nearly two thirds of the total poor in the world. In 1998, 40% of the people in South Asia alone lived on less than a dollar a day, and around 15% in East Asia and the Pacific. The human development values and gender development indexes for South Asia, East Asia, and Asia and the Pacific also show substantial variations in the region, with the lowest levels of achievement in South Asia. Urgent strategic energy measures are needed to address these challenging situations, and help move women into modern energy systems (Wickramasinghe 2007a).

The energy system uses different types of fuel, equipment and supply sectors to deliver services such as illumination, cooked food, heating, refrigeration, communications, education, transportation and motive power. Gender roles in energy systems depend on the type of energy being considered, available livelihood opportunities, the nature of the assets involved and the social status of women (Table 1). In countries where a substantial share of primary energy is supplied from biomass such as in Sri Lanka, Vietnam,

Table1
Roles of Women in the Energy System by Major Types of Energy in Asia

Energy type	*Roles of women*	*Status of women*
Biomass	• Maintaining vegetation cover through farming/ growing tree crops • Harvesting, gathering, processing and transporting wood in bulk • Converting raw wood into heat through combustion • Providing energy services – energy for cooking, boiling water, preparing medicine for home remedies, heating, processing food and crops; • Managing biomass fuel stocks at the household level to even out variations in availability; • Maintaining traditional end use technology – cook stoves and earthen hearths/ kilns.	• Social sector/ key stakeholders responsible for all upstream activities, conversion of primary energy to useful energy and providers of energy services • Resource less customary users • Unpaid family labour source • Unauthorised access to sources and management.
Electricity	• Providing primary sources for electricity generation (dendro, biogas) • Managing / contributing to renewable energy resources • Undertaking the responsibility of sustainable management of technology	• Consumers of energy services (cooking, lighting, communication, mechanisation, water pumping) • Managers of off-grid electricity supply plants • Custodians of decentralised energy supply system • Managers of primary energy sources • Energy entrepreneurs

Contd.

Energy type	Roles of women	Status of women
	• Sustaining primary energy supply chain for electricity generation • Contributing to conservation of sources and resources	• Partners of energy organisations • Organisers of local demand for energy carriers
Fossil fuel	• Managing production (SMEs) and cottage based industry • Adopting energy efficient technologies and appliances • Promoting micro-enterprise and enhancing market economy • Reducing wastage.	• Consumers of energy services (transportation, lighting, water pumping, irrigating, grinding, processing, industry etc.) • Organisers of local demand for energy carriers

Source: Wickramasinghe (2007a)

Bangladesh, Myanmar and Nepal – biomass is an affordable and accessible source of energy. Women's involvement in the biomass system from production to service generation is also high. If women can retain their energy roles using technology through which biomass feedstock can be transformed to modern energy carriers with commercial opportunities, this would be a win-win situation. Economic empowerment could be promoted by creating economic opportunities for women using modern energy technology, while enhancing women's capacity to access energy services through income generation. The synergy between energy and all aspects of development–social, economic and environmental including poverty, hunger, education, and gender, has been demonstrated, and there is wider understanding about the importance of equal access to affordable, clean and modern energy services for sustainable development.

Bangladesh

Due to its agriculture-based economy, Bangladesh has a very low rate of per capita commercial energy consumption compared with neighbouring countries in South Asia. Non-commercial energy sources, such as wood, crop residues and animal wastes, account for half of the country's energy consumption (Khan 2007). There are indications that consumption of biomass energy has already exceeded the regenerative limit, which will lead to an ecological crisis in rural areas of Bangladesh. Like many other developing countries, the energy sector of Bangladesh remains centralised within the government, reflecting its importance to the economic development of the country. The policy planning of the energy sector is influenced primarily by large industries, and closed to input from stakeholders and advocacy groups representing households and small businesses. Procedures for setting national energy priorities, therefore, do not often take into account the differences between men and women in energy usage or needs. This is most apparent when it comes to the rural energy sector. Despite their key roles in energy collection, purchase and usage, women are not generally targeted as primary beneficiaries of energy services. Resolving these problems and improving the status of women will require an understanding of gender relations, as well as gender-focused and gender-mainstreaming actions.

In Bangladesh, women face inequality in the socio-economic and cultural structure. Their family responsibilities limit their access to the labour market, their gender limits their choice of occupations, and within paid jobs their wages are not equal to those of men. Most women, including not just housewives, but also entrepreneurs, advocates and energy professionals, currently have little role to play in energy decision-making. In the typical work environment at energy-related enterprises and industries, women professionals tend to be segregated into administrative and accounting job categories, whereas men are more often in the technical and field-based jobs. Furthermore, women in Bangladesh have often joined energy institutions later than men, which contribute to the fact that they are not well represented at the senior levels of management. On the Rural Electrification Board, for example, approximately 8 per cent of the officers are women, and only 11 per cent of the total workforce. In the rural electrification cooperatives, only 20 per cent of the staff is women (ENERGIA and ECBL 2003)).

The absence of women in energy programming and an overall lack of recognition of energy- related gender issues have several important consequences. As energy service providers remain out of touch with household and small business energy needs, and energy providers and users are oblivious to sustainable energy alternatives, there is an increasing likelihood of environmental degradation from continued reliance on traditional biomass fuel collection and combustion. Wood supplies are being depleted with the increasing demand of fuel, and agricultural productivity is affected by removal of organic matter from the fields.

India

In India the main sources of energy are currently fossil fuels and biomass, structural adjustments in the energy sector aim towards privatisation, without addressing the need for energy access for the poorer parts of the population. For women, there are additional barriers of a social, political, economical and cultural type, when compared to men, which impede their development on terms of equality. Only on a few occasions do they find themselves in positions that allow participation in decision-making with respect to programmes or projects in the field of energy, while generally in rural areas the women are the ones dedicating their time to collecting fuel wood to prepare food for all the family, making them the main consumers of household energy.

Both rural and urban Indian women need adequate energy supplies for their small and medium-scale enterprises and home industries. Many of these informal sector activities are highly fuel-intensive, and their viability and costs are affected by energy prices and availabilities. Micro-enterprise development initiatives, such as self-help groups, should be emphasised, interlinking micro-credit and energy programmes, and providing promotional incentives for running small-scale energy business units to enhance the employment opportunities for women and encourage them to use resources available locally. The role of women as energy providers can be transformed into suitable micro-enterprises if they can manage fuel wood or oil seed plantations, dispense kerosene or LPG, assemble solar panels, build cook stoves and brick kilns and even manage electricity distribution and bill collection. New opportunities for sustainable technologies can be more readily adopted if women have the necessary resources to participate in the formulation and implementation of energy programmes and policies (Sarkar 2006b).

Nepal

About 84 per cent of Nepal's population lives in rural areas, with agriculture as their main livelihood. For the period 2003/04, total rural energy consumption was 288 million GJ, of which 97 per cent was residential: cooking, 63.9 per cent; heating, 8.5 per cent; lighting, 1.31 per cent; agro-processing, 3.4 per cent; animal feed preparation, 16.5 per cent; and others uses, such as for religious occasions and ceremonies, 4.3 per cent. Of the total energy consumption of 288 million GJ in rural Nepal, biomass accounted for 98 per cent. Electricity accounted for only 0.1 per cent of the total energy consumed, petroleum products comprised 1.6 per cent, and renewable energy sources 0.5 per cent (Bhadra and Shakya 2007).

Currently, the household maintenance burden lies on women's shoulders in poor households, in part due to the male out-migration. In order to reduce poverty in Nepal it is an imperative that women at the grassroots level are economically empowered through energy interventions fostering income generation and enhancing agricultural production. For this, it is recommended that there be 'hybridisation' of rural energy development initiatives with income-generation plans to ensure affordability for poor women. As rural energy development is costly, the fear that women will not be able to pay for it generally keeps women from accessing rural energy technology. However the 'hybridisation' of energy technology with income generation opportunities for women ensures their ability to pay. Examples of income-producing enterprises by women's groups that could be linked to energy technologies include food processing mills, power looms, forest nurseries, poultry farming and production of beehive briquettes. For this, the energy sector policies need to remove the currently used demarcation between energy for domestic use and energy for enterprises and farming. This way, the energy intervention with a gender perspective will take care of both women's 'practical gender needs' of reducing time-use and drudgery in household and agricultural works and enhancing agricultural productivity, as well as taking care of women's 'strategic gender interest' in economic empowerment through income generation. To ensure energy-related enterprise ownership by women, financing of rural energy development need to be prioritised, with channelling of credit to women's groups matched with managerial capacity building. Furthermore, it is very urgent to expand the coverage of renewable energy technology to include a larger number of poor households if an immediate and effective impact on poverty and hunger reduction is to take place. Biomass (especially firewood) is the energy of the poor but this is also a source that is diminishing valuable forests, emitting indoor air pollution, and accelerating adverse climate changes globally. Indoor pollution impacts on women's health, and climate change has affected women farmers' systems and productivity. Access to and utilisation of alternative energy technologies such as biogas and improved cook stoves, and improved water-mills for agro-processing and lighting, directly reduces dependency on the forests, resulting in increases in forested areas, which could be protected for maintaining biodiversity. Similarly, micro-hydro and solar energy can reduce imports of fossil fuels and reduce greenhouse gas emissions, as well as saving valuable foreign currency. Additionally, such clean technologies traded in the Clean Development Mechanism (CDM) market (established under the Kyoto Protocol) can earn valuable foreign currency. The saved money through commercial energy substitution and money earned in the CDM market can be used for necessary development activities in the country, especially research and development targeted to women-friendly energy technologies.

Sri Lanka

In 2002, biomass provided about 50% of the primary energy, about 84% of which was for domestic and commercial uses, and the rest for the industrial sector. Nearly 90% of the population uses biomass in the form of fuel wood and agro-residues for cooking. The annual consumption of fuel wood is nearly 10 million tonnes, supplied primarily from non-forest production sources. Around 26% is provided from home gardens, 19% from coconut plantations, 19% from croplands, 7% from natural forests, 7% from rubber plantations, 4% from forest plantations, 3% as processing residues and the remaining 14% from other sources. Lack of access to affordable and modern energy services for the poor, and particularly for women, directly and indirectly affects the country's progress in achieving sustainable development goals and gender equality. Sri Lanka has not been successful in promoting women's equal access to and participation in decision-making at all levels on the basis of equality with men (Wickramasinghe 2007b).

Disparities with regard to gender involvement in making decisions on clean energy supply, technology, use and management are quite wide. Women's lack of capacity to secure and spend money on clean energy leads them to spend their productive labour on gathering and portaging fuel wood from freely available sources, despite the detrimental effects on their health and the economy. Women lack the financial capacity to get access to energy technologies for lighting, household appliances, motorised pumps, and equipment for grinding and crushing grains. Enhanced access to transportation is also essential. The conditions under which women continue their roles as providers of biomass for generating cooking energy are critical to issues of national importance, such as management of energy resources, ownership of natural resources, distribution of fuels, technology choices, and housing construction, and have an adverse effect on many other aspects of women's lives, like income, nutrition, and time availability. Lack of investments or financial services for women is a result of men's dominance in decisions about energy enterprises. Financing for women to gain access to clean energy, and securing modern energy for women's productive enterprises are equally important. Energy is needed for promoting the use of information technology and creating new partnerships and opportunities, especially for women in disadvantaged locations and positions. The socialisation and localisation of energy technology suffers from lack of gender sensitivity. Gender conscious energy technology planning and integrating of financial services for promoting technology and its extension is essential.

Vietnam

Within the energy sector, women are producers, processors and also important end users. Approximately 60% of fuel gatherers are women. They are also the first ones to bear the burdens of energy shortages. Women should be the main focus for energy conservation activities, since their work consumes most of the energy used by the family. However, even when they are aware of energy options, they have little access to financing for better equipment. Traditional social norms have determined that a bigger portion of the financial decisions are taken by men than by women, as it is thought that greater results can be expected from men. Activities such as negotiating with banks and credit organisations are considered better suited for men. Vietnamese women tend to meet more difficulties in accessing official credit sources. They generally have to use unofficial lenders offering higher interest rates and limited amounts of money. Accessing any kind of loan is difficult for poor people, and the prospects are worse for poor women. Given this situation, women find it hard to have enough money to cover the costs of buying new energy-saving tools or applying energy-conservative technologies. Less participation in training and education in technical areas keep women away from potential high-tech jobs, including ones in energy fields. As a result, there are few women participating in drafting energy policies or contributing to the energy conservation process (Anh 2007).

Indonesia

Energy demand in Indonesia has sharply increased due to national development and population growth. Over the last three decades, total commercial primary energy consumption rose more than tenfold, from about 50 million barrels of oil equivalent (BOE) in 1970 to about 760 million BOE in 2003. This increased commercial energy demand has been met mainly by oil (52%), natural gas (21%) and coal (20%). Only 7% has been from renewable energy, mainly geothermal and hydro. However, the largest proportion of energy consumption in Indonesia is still for household use. According to a survey conducted by the University of Gadjah Mada, household activities account for more than 40% of the country's total energy use. Due to traditional gender roles, supplies and utilisation of household energy for cooking and home appliances are primarily managed by women rather than men (Utami 2007).

The economic condition of women is one of the most crucial aspects of women's welfare. The energy sector could improve women's economic empowerment in several ways. Energy-based interventions could contribute to lightening the burdens of women, in a social context where women expend a large portion of their time and strength on energy-related household tasks. Through interventions improving energy services and enhancing the effectiveness and efficiency of energy devices related to women's household tasks, women could also gain more free time to pursue training and income generating activities. For example, improved cooking devices would benefit women not only in feeding their families, but also in home food production and other related businesses that make up primary areas of women's income-producing work. For these sorts of interventions, there should also be a strategic selection in terms of target areas and target populations; prioritisation should be directed as much as possible towards areas and populations selected based on consideration of the human poverty index, gender related development index and human development index.

Lao PDR

The energy situation in Lao PDR is characterised by low conventional energy consumption. Biomass accounted for 78% of the total energy consumption in 2002, followed by petroleum products (16%), and electricity (6%). The high proportion of fuel wood consumption reflects its widespread use, particularly in the rural areas. Imported oil and LPG are mainly used in the industrial sector. Wood fuels consumption in 2002 was 2.4 million tons and accounted for 69% of total energy consumption in Lao PDR. It is estimated that about 94% of the households use wood fuels for cooking. Charcoal is one of the important traditional sources of energy for both urban and semi-urban populations, and is used mainly for cooking. Most wood energy users in Laos are women. As part of their activities, they are often responsible for gathering fuel wood, and doing the cooking. Many of the people who are involved in the wood fuel trade or who work in rural industries or commercial enterprises that use wood fuels, are women. This means that in wood energy, gender issues play an important role. There is a serious lack of skilled female staff in the energy sector, especially technicians. Women's technical skills are often less than men's, and women in the energy industry work mainly in administration, sales, finance, and secretary. Traditionally Lao women have played a key role in energy management - in terms of collection and usage at the household level. Past experience in Lao PDR has shown that most of the energy-related interventions have not mainstreamed gender concerns in their projects (Siliphong 2007).

Gender Sensitivity in Meeting Energy Needs in Africa

In Africa, women are responsible for the provision of products, labour and services as part of their household and family obligations, reciprocal relationships and mutual support. This is usually done as unpaid domestic work, which is often treated as of no value in national accounts. Decision-making is largely in the hands of men, who assume that the unpaid work of women is free and forever available. Thus bearing and raising of children, preparing and growing food for the household, and gathering fuel and water appears invisible and insignificant in the understanding of economic development. Yet a lot of labour is required to provide the energy needed to meet these responsibilities. Women therefore find themselves having to bear a disproportionate share of the physical, emotional and spiritual strain of providing such energy resources.

In Africa, women and girls spend long hours every day collecting wood, agricultural residues and dung for use as fuel - time that could be used for more productive activities such as income generation, agricultural production or education. In this situation women and girls are likely to benefit most when there is increased access to improved energy services. There are, however, variations among countries as time spent gathering biomass depends on geographical location, land ownership, time of year, climatic events and loss of control over local resources.

There is a need for liquefied petroleum gas (LPG) and other cleaner fuels to provide affordable alternatives to traditional biomass-based cooking and heating fuels, but the development of advanced and cleaner energy technologies and fuels has generally not been perceived as a high priority in Africa. In some countries the rate of access to LPG, for example, is lower than 5% (ECOWAS 2005). Senegal has been a pioneer in the region in terms of the efforts it has made to make butane affordable and easily available, at least in urban areas. But generally, women in rural and peri-urban areas in Africa are compelled to use charcoal and fuel wood (ECOWAS 2005).

In Africa, regional commitments, targets and goals to address energy concerns include energy accessibility for poverty alleviation; changing patterns of energy consumption and production; developing advanced and clearer energy technologies; and general cross-cutting issues. Energy policies and projects have tended to be conceived and implemented from a gender blind perspective, despite the many international agreements and conventions that African governments have signed to promote sustainable development, gender equality and environmental protection. Energy planning is generally done using a top-down approach, seeking to increase supplies of fuel or electricity for industrial and urban uses, with low interest in women's specific energy needs and concerns. Energy sector restructuring exercises have tended to focus primarily on cost-recovery measures, regulatory and legal frameworks, privatisation, and technical problems (Modi *et al.*, 2006). While some of such measures are useful, their uncritical application and the lack of sufficient interest in promoting gender concerns have proved problematic. They have tended to orient energy initiatives towards profit making, rather than promoting easy and affordable access, improving reliability for current users, and expanding services to benefit new customers in line with MDG goals and objectives. Thus areas such as small-scale informal productive activities, domestic tasks and agricultural work where women actively participate are not factored into many national and local initiatives. A major underlying cause for this state of affairs is that energy policies in Africa have been formulated as part of overall economic policy frameworks, which have proved ineffective in addressing the development concerns of women and men. These policies are rooted in the broad neo-liberal economic frameworks that have been implemented since the 1980s. Such policies have had negative implications for women, ignoring prevailing socio-economic realities, laws, customs and practices in individual countries that often restrict or prevent women from actively participating in decision-making or having independent access to control of income and ownership of property. The overriding interest in promoting specific energy technologies has led to a neglect of the gender specific inequalities in access to and control of affordable energy services. Thus within the region, in spite of the positive impacts on women of some energy initiatives, there is no indication that energy policies, programmes and projects are creating opportunities for transforming the gendered relationships between women and men, with a view to reversing years of discrimination against women. To accomplish this, energy policies must not only target women, they must actively seek the participation of women in making the appropriate decisions on enhanced energy service.

The lack of sufficient participation of women in decision making processes generally, and energy sector decision-making specifically, is part of the problem, although the African region as a whole has made important strides in women's participation in politics and decision-making. Through the use of quota systems, the participation of women on the African continent in the past few years has been greater than that experienced at any other time in the past four decades, rising more than tenfold to over 14% in 2003. Over 20 countries in the region either have legislated quotas or political parties that have adopted them voluntarily (IDEA 2004). These measures have contributed directly to the increase in the number of women in legislatures. In 1995, the average representation of women in sub-Saharan Africa was 9.8 per cent; it had increased to 15.1 per cent by 2004 (IDEA 2004). There are clear disparities in different countries, with Rwanda topping the list worldwide. There are also variations in representation in various sectors, with the energy sector being among those with minimal representation. There are situations where women's participation in decision- making becomes flawed due to the fact that some women tend to represent sectional interests rather than the majority of women. There is also no guarantee that women decision-makers will address energy issues from a gender perspective, as there is often insufficient understanding of the issues on the part of women leaders. There are also a low proportion of women professionals in the energy sector, even in situations where affirmative action policies have been implemented to increase women's participation in decision-making across the board (Wamukonya 2002).

There have been attempts at increasing women's participation in energy decision-making through the establishment of gender desks across energy ministries, departments and agencies. These are often staffed by women who are expected to integrate gender considerations into policies and planning. The reality, however, is that such offices are often isolated and have few linkages with other ministries and critical institutions. Such structures, which are similar to national level institutional mechanisms for women, also lack the needed budgets and mainly rely on irregular donor funding, which makes the work unsustainable (Mama 2000).

Botswana

Fuel wood accounts for 43% of the final energy consumption in Botswana, while electricity accounts for only 11%. The degree to which fuel wood is used in rural areas for cooking and water heating is 78% and 83%, respectively, as compared to urban areas where it is about 17% and 32%, respectively. Access to electricity is low (31%) in rural areas. By 2004, a total of 238 out of 400 villages were electrified in the country. In view of the low rate of access to electricity, the government has intensified its rural electrification programme, using both grid-based and off-grid sources (Botswana Technology Centre 2006).

Botswana has made progress, albeit slowly, to include women in all facets of economic development. The slow progress in enabling women to hold top decision-making positions continues to keep women's needs, including their energy needs, inadequately addressed. It is tricky for men to articulate women's energy requirements, or create energy technologies and provide energy services to be used by women, without consultation with women or without having women involved in the policy formulation process. Programmes and policies, which lack women's involvement, are likely to misrepresent their needs. Gender mainstreaming is another issue that needs to be addressed. There is general awareness regarding the different roles of women and men in the country and men are slowly accepting that there are gender differences, which must be considered when policies are formulated. The Women's Affairs Department has made some progress in mainstreaming gender in some sectors but this remains inadequate, and as a result women's development is lagging. The department is understaffed and is therefore unable to mainstream gender in all the sectors as a matter of urgency. This is particularly the case in the energy sector where many women are affected negatively by policies due to lack of capacity by policy-makers to appreciate the impact of energy services on the different household demand groups, i.e., men and women.

Ghana

There is a critical link between energy and economic activity, and the impacts one on another is seen not only the livelihoods of people, but also on several other drivers of sustainable development including water, agriculture, and health. With a rural population of over 11.4 million people and a human development value of 33.1 per cent, Ghana ranks 58th among 102 developing countries on poverty levels (UNDP 2005). In this context, inequitable access to clean energy sources in the country is major impediment to sustainable development. This is amply borne out by the fact that misery of close to 37 per cent of the rural masses who do not have access to safe source of drinking water and over 58 per cent who lack access to proper health care can be addressed to a large extent through the provision of clean and efficient energy sources (Ghana Statistical Service 2005).

In terms of energy demand, the household sector is the largest consumer of energy, followed by transport and industry. As for energy supply, only one-third of Ghana's total energy supply is in the form of commercial energy, while the rest is traditional fuels, which are low efficiency and mainly burned for cooking. Key energy issues in Ghana are the unavailability or inaccessibility of modern energy sources to a large fraction of the population, non-affordability of commercial fuels, and high-energy import dependence.

Women are the most important actors in the energy sector in Ghana in terms of their contact, use and management of renewable energy sources, which in their very crude or primary form are used mostly by women. The most common energy sources are fuel wood, charcoal and cow dung. Fuel wood and charcoal account for more than 75 per cent of all energy requirements in the country, and an even higher percentage of energy for household cooking and water heating – in rural and urban areas alike (Owusu *et al.*, 1989).

Despite the enormous responsibilities and risks women bear and are exposed to in their bid to ensure affordable energy sources, and their overall importance in the energy sector, their views are rarely sought in the decisions taken on issues concerning energy. This is basically because most of the decision-makers, managers and technical personnel are men and their attention is given more to generating more fossil fuel and electrical energy for industrial purposes than for domestic purposes that would show some level of sensitivity to women's energy needs and improve their situation. Probably also because of economic returns that fuels used for commercial purposes bring in, any appreciable attempt at according fuel for domestic purposes is thought of as being a big waste rather than an investment. Despite all these economic calculations, it behooves the state and all its institutions and relevant agencies to see to it that investments are also made in the economically viable things that will lead to an improvement in the lives of the majority of its citizenry, which happen to be poor rural and urban woman.

Lesotho

Lesotho faces significant challenges with regard to energy for sustainable development. About 85 to 90 per cent of the country's total energy consumption is still utilised in the domestic sector, and is used for cooking, heating water and space, lighting, and powering electrical appliances. Sources of energy include electricity, liquefied petroleum gas (LPG), paraffin, firewood, coal, biomass fuels (i.e., crop residues, dung, fuel wood and shrubs), candles and solar energy. These sources and their extent of use differ between urban and rural households, depending on the availability of the source, and its cost vis-à-vis household income. Commercial energy constitutes about 10 to 15 per cent of the country's energy demand, and is provided through imports from South Africa (Redeby 2007).

Women's organisations should emphasise the prioritisation of sustainable energy specifically targeted at women's energy needs over conventional sources of energy, which are not particularly suited to the energy needs of women in Lesotho. This can help contribute to the achievement of the country's MDGs, in that women's energy needs will be more adequately addressed, and the serious burdens faced by women in their duality of work and family roles will be alleviated. Provision of modern energy services can free women from their time consuming and marginally productive activity of sourcing energy, and instead provide them with more time for income generating and poverty alleviating activities. Growth and employment opportunities could thus be generated in sectors where the poorest people, predominantly women, are concentrated, such as agriculture, industry and small and medium-scale enterprises. This will help to achieve one of the most important MDGs for Lesotho, which is eradication of extreme poverty and hunger.

Since burdens from collecting biomass fuel and lack of alternative energy fall so heavily on women in Lesotho, there is an urgent need to assess energy programmes in terms of their effects and benefits for women and involve women in decisions regarding both the planning and implementation of energy solutions. It is particularly important to coordinate energy policies with broader economic, health and poverty reduction efforts, including specific initiatives to continue to expand economic opportunities for women.

Nigeria

Nigeria's energy policies and programmes have generally not been gender sensitive. Information about disaggregated patterns of energy use between men and women in households are not available to decision-makers and planners, largely due to lack of concern and understanding about gender issues. However, it is clear that with regard to household energy, fuel wood is the pivot on which the domestic life and the economic activities of rural Nigerians are based, and that women are the major users of fuel wood. Women use up a lot of physical energy walking long distances searching for ever-depleting supplies of needed fuel wood.

However, the 2003 Nigerian Energy Policy adopted the traditional approach to energy in development by assuming that energy is gender neutral, and thus failed to address women's particular energy needs. There remains a need to devise the type of energy policy that will address the present and the future energy needs of the urban poor, women and children, and rural populations.

One of the major energy policy constraints identified in the Nigerian Sustainable Development policy document is inadequate local capacity in energy planning and management.

To empower women and engender energy for sustainable development, there needs to be increased investment in local capacity building and technological education, especially for women (Chikwendu 2007).

South Africa

South Africa has large supplies of coal reserves, which provide 70 per cent of the country's primary energy. South Africa also generates over half the electricity used on the African continent, mostly from coal-powered generating plants. There are also large reserves of uranium, and smaller reserves of crude oil and natural gas. South Africa imports the bulk of its oil requirements.

With regard to renewable energy, there is limited potential for hydropower due to the country's arid terrain. However, there is fairly good potential for development of solar and wind energy. Biomass is a critical energy source for households, especially in rural areas, yet the potential for sustainable use of biomass as fuel is not good, again because of the arid nature of the country. Commercially, energy from biomass is used in the sugar, refining and pulp industries, and there is also some potential for production of biomass fuels, such as ethanol.

Cooking, ironing, and water heating are some of the women's main tasks - in fact, they are often the sole responsibility of women. The procurement and management of energy sources required for these tasks then also falls on women. However, due to relationships of power and control, as well as economic constraints, women don't often have the ability to choose the energy sources they would like to use. Because energy services requiring thermal energy are viewed as women's jobs, the subject has never been adequately addressed, or appropriate solutions to problems sought. Women in poorer households have little choice and must use whatever fuels are available and affordable, even if they are dirty, dangerous and inconvenient. Even when electricity is available, most households continue to use paraffin, wood or gas for cooking (Hooper-Box 1998).

CONCLUSION

In many developing countries, especially in the poorest areas, most energy currently comes from traditional biomass fuels such as wood, charcoal and agricultural wastes and collecting and managing these fuels is strictly 'women's business'. In order to expand access to energy most effectively, in ways that promote economic and social development, it is important to determine which types of energy services are needed most, and which ones will best support new income-producing enterprises. Gender sensitivity in energy policies is not just a matter of taking women's feelings into account. It involves acceptance of existing economic realities for women, and new financial investments in energy infrastructure and services that will best promote economic and social development for both men and women. Expanded investments in energy are critical for achievement of the internationally accepted Millennium Development Goals (MDGs), since without better access to energy, many people in developing countries have little chance of escaping from lives of extreme poverty. It is also time for greater attention to gender in energy investments and initiatives. The goal is not to remove women from the energy sector, but actually to involve them more, and in different ways, so that they can manage their energy supplies, their businesses and their lives more effectively and productively.

Energy policies and interventions that recognise women's roles in the energy sector and build on their expertise and influence will be much more effective in promoting sustainable economic and social progress. In places where traditional biomass fuels are collected and managed primarily by women, there are serious economic and social consequences. These activities substantially increase women's physical burdens, damage their health, and take up time that could otherwise be spent on caring for their families, educating themselves and their children, and engaging in income-producing activities. Burning biomass fuels is also a major contributing factor in high sickness and mortality rates for women and children.

Many strategies and approaches have created some enabling conditions, if not a transitional move, towards integrating gender and energy into the development paradigm. Mainstreaming gender in energy activities has provided windows of opportunity enabling women to gain greater access to modern energy and services to meet their development needs and affirm their equality.

In many Asian countries, an integrated village development approach, which includes energy, has made significant changes in the lives of women. The need for providing a wide range of energy options for satisfying women's energy needs has been well demonstrated. The energy types used in generating energy services is rather diverse; access, affordability as well as the economic feasibility and efficiency have been identified as key concerns influencing energy choices.

Within the African context, policies of privatisation in the energy sector have formed part of the economic reform agenda. The involvement of private companies as major players in the energy sector means that the profit motive is prioritised. This means that rural and urban poor women are not served in many situations. But for the majority of the unserved urban and rural poor women of Africa, a major concerted effort by all African governments is required, with effective support from the international community, to ensure that modern energy services are easily accessible and affordable, and promote improved relationships between women and men on an equal basis.

The energy contributions and needs of women have therefore remained largely unrecognised in most of the countries national policies, planning processes and development assistance efforts. Perhaps it seems somewhat embarrassing for governments to admit that poor, uneducated women dominate the country's energy sector. Nevertheless, government officials need to acknowledge that in many places the energy business is in fact women's business, and build on that base rather than ignoring or rejecting it.

REFERENCES

Anh D. Q. (2007), Gender and Energy in Vietnam, In: Gail Karlsson (Ed.), *Where Energy is Women Business,* ENERGIA, Leusden.

Bhadra C., and Shakya I. (2007), Nepal: Prospects and Challenges of Gender Mainstreaming in the Energy Sector, In: Gail Karlsson (Ed.), *Where Energy is Women Business,* ENERGIA, Leusden.

Botswana Technology Centre (2006), *Gender Audit of Energy Policies and Programmes: The Case Study of Botswana,* Gaborone.

Chikwendu C. (2007), Engendering Nigerian Energy Policy, In: Gail Karlsson (Ed.), *Where Energy is Women Business,* ENERGIA, Leusden.

ECOWAS (2005), *White Paper for A Regional Policy,* UNDP and French Ministry of Foreign Affairs.

ENERGIA and ECBL (2003), Comprehensive Gender Strategy and Action Plan for Integration in REB Master Plan, Royal Netherlands Embassy, Dhaka.

Energy Sector Management Assistance Programme, (ESMAP) (2005), *Advancing Bioenergy for Sustainable Development: Guideline for Policymakers and Investors, Volumes I, II and III,* Energy and Water Department, World Bank, Washington DC.

Ghana Statistical Service (2005), *Ghana 2003 Core Welfare Indicator Questionnaire (CWIQ II) Survey Report,* Accra.

Hooper-Box C., *et al.* (1998), *A Conceptual Review and Reassessment of South African Energy Policy Research Methodology: Towards an Expanded Framework,* Department of Minerals and Energy, Republic of South Africa.

IDEA (2004), *The Implementation of Quotas: African Experiences,* Trydelis and Trycheris AB, Stockholm.

Khan H. J. (2007), Energy, Gender and Achievement of the Millennium Development Goals in Bangladesh, In: Gail Karlsson (Ed.), *Where Energy is Women Business,* ENERGIA, Leusden.

Mama A. (2000), *National Machinery for Women in Africa: Towards an Analysis,* National Machinery Series No. 1,Third World Network Africa Secretariat, Accra.

Modi M., *et al.* (2006), *Energy and the Millennium Development Goals,* Energy Sector Management Assistance Programme, UNDP and World Bank.

Owusu M., *et al.* (1989), *Report of the Working Group on Forestry and Wildlife, Revised Version*, Prepared for the Environmental Protection Council, Accra.

Redeby L. K. (2007), Mainstreaming Gender and Energy in Lesotho, In: Gail Karlsson (Ed.), *Where Energy is Women Business*, ENERGIA, Leusden.

Sarkar S. (2006a), Cooking with Biofuels and Women Health, *International Journal of Women, Social Justice and Human Rights*, 1(2): 161-184.

Sarkar S. (2006b), Energy Security for Rural Poor, *Journal of Social and Economic Policy*, 3(2): 235-244.

Siliphong P. (2007), Gender and Energy in Lao PDR, In: Gail Karlsson (Ed.), *Where Energy is Women Business*, ENERGIA, Leusden.

UNDP (2004), Energy for Sustainable Development in Asia and the Pacific Region: Challenges and Lessons from UNDP Projects, New York.

UNDP (2005), *Human Development Report. International Cooperation at a Crossroads: Aid, Trade and Security in an Unequal World*, New York.

Utami P. (2007), Gender Sensitivity in Meeting Energy Needs to Overcome Poverty in Indonesia. In: Gail Karlsson (Ed.), *Where Energy is Women Business*, ENERGIA, Leusden.

Wamukonya N. (2002), *A Critical Look at Gender and Energy Mainstreaming in Africa*, Available at *www.un.org/womenwatch/daw/forum-sustdev/Njeri-paper.pdf*

Wickramasinghe A. (2005), *Annotated Bibliography: Gender and Energy in Asia*, ENERGIA, Leusden and National Network on Gender and Energy, Peradeniya.

Wickramasinghe A. (2007a), Gender and Energy in Asia Regional Initiatives and Progress in Mainstreaming Gender and Energy, In: Gail Karlsson (Ed.), *Where Energy is Women Business*, ENERGIA, Leusden.

Wickramasinghe A. (2007b), Gender and Energy Issues Related to Sustainable Development: Sri Lanka, In: Gail Karlsson (Ed.), *Where Energy is Women Business*, ENERGIA, Leusden.

18

LAND USE/LAND COVER MAPPING USING DIGITAL IMAGE PROCESSING OF IRS-ID SATELLTE DATA: A CASE STUDY OF MORAR WATERSHED, DISTRICT GWALIOR M.P. INDIA

Prafull K. Singh & U. C. Singh

INTRODUCTION

To meet the demands of large population means the need for more food production, more requirement of energy, more water requirement, better civic amenities for a reasonable quality of life, more infrastructure development to sustain increasing pressure and increase per capita expenditure for maintaining quality of life. This requires the utilization of land use/land cover analysis in the area.

Land use/land cover mapping is an essential component wherein other parameters are integrated on the requirement basis to derive various developmental index for land and water resources. In order to undertake a proper, systematic and structured land cover/ landuse mapping, it is important to identify land cover/landuse classes as per NRSA (1989) classification of land use/ land cover scheme. The unparallel increase in population growth has resulted in the increasing demands for food, fodder and shelter resources. This calls for rational utilization of the available land and hence the study of its characteristics is crucial in formulating the management and developmental plans. Land use is a product of interaction between a society, cultural background and its physical needs on the one hand, and the natural potential of land the on other (Balak *et al.*, 1993).

The knowledge of spatial land use/land cover information for the proper management, planning and monitoring of the natural resources are very essential. There are several urban applications where satellite based remotely sensed data are being applied, namely, urban sprawl/ urban growth trends, mapping and monitoring land use/land cover, urban change detection and updation, urban utility and infrastructure planning, urban land use zoning, urban environment and impact assessment, urban hydrology, urban management and modeling. Rao (1991), Chaurasia *et al.*, (1996), Pathak *et al.*, (1998), Srivastva *et al.*, (2003), Balaselvakumar (2003) Shaha *et al.*, (2005) and Chaudhary *et al.*, (2008) have used the remote sensing and GIS based approach to prepare land use/land cover maps of various terrains. Remote sensing with its advantage of varying spatial and spectral resolution and temporal availability of data covering large and inaccessible areas within short time has become a very handy tool in assessing, monitoring and mapping the land use/land cover resources of the area.

The main objective of the present study was to evaluate the present status of land use/land cover in the watershed.

STUDY AREA

The Morar Watershed lying between the latitudes 26°5¢-26°25¢ N and longitudes 78°10¢-78° 25¢ E and covering an area around 405 sq. km. in Gwalior district of Madhya Pradesh has been selected for the present study. It is covered in the Survey of India topographic sheet nos. 54 J/3, 54 J/4, 54 J/7and 54J/8 Gwalior district is surrounded by the districts of Morena in the north, Shivpuri in the southwest, Bhind in the northeast and Datia in the southeast. It is located (Figure 1) on the junction of National Highway No. 3 and 75 and on the tracks of North Central Railways. It is about 1125 km. northeast of Mumbai, 318 km. south of Delhi and 388 km. north of Bhopal. Vijaya Raje Scindia airport is located in the northeast of Gwalior city near Maharajpura village; about 20 km. from Gwalior Railway Station. Within the watershed area under investigation, there exist Air Force Station at Maharajpura having the Miraj Fighter Plane Air base and Army Station Headquarter at Morar which make the city strategically important at the national level. The area experiences a semi arid climate marked by extreme temperature and variability of rainfall. The duration of winter season is from November to February followed by the summer from March to middle of June. The rainy season starts from middle of June to September when southeast monsoon is active while October and November are post monsoon or retreating monsoon month.

GEOLOGY

Geologically, the area forms the part of terrain comprising hard rocks ranging in age from Archaean to Recent. Bundelkhand granite complex comprising pink and gray granites with enclaves of basic and ultrabasic rocks are exposed in the southern part. The veins of pegmatite, quartz and basic dykes trending NE-SW and NW-SE traverse these granitic rocks. The overlying Gwalior group is divided into two formations: the older Par formation and younger Morar formation covering northeastern and the central parts of the district with dolerite dykes present only in Morar shale (Figure 2) which reveals that Morar is younger than Par. This group of rocks are overlain by upper Vindhyans (Kaimur sandstone) which are the youngest in the region (Pateriya and Singh, 1994).

DATA USED AND METHODOLOGY

In the present investigation, IRS ID LISS-III and PAN digital data (Path 97 and Rows 53) dated 22 February, 2003 along

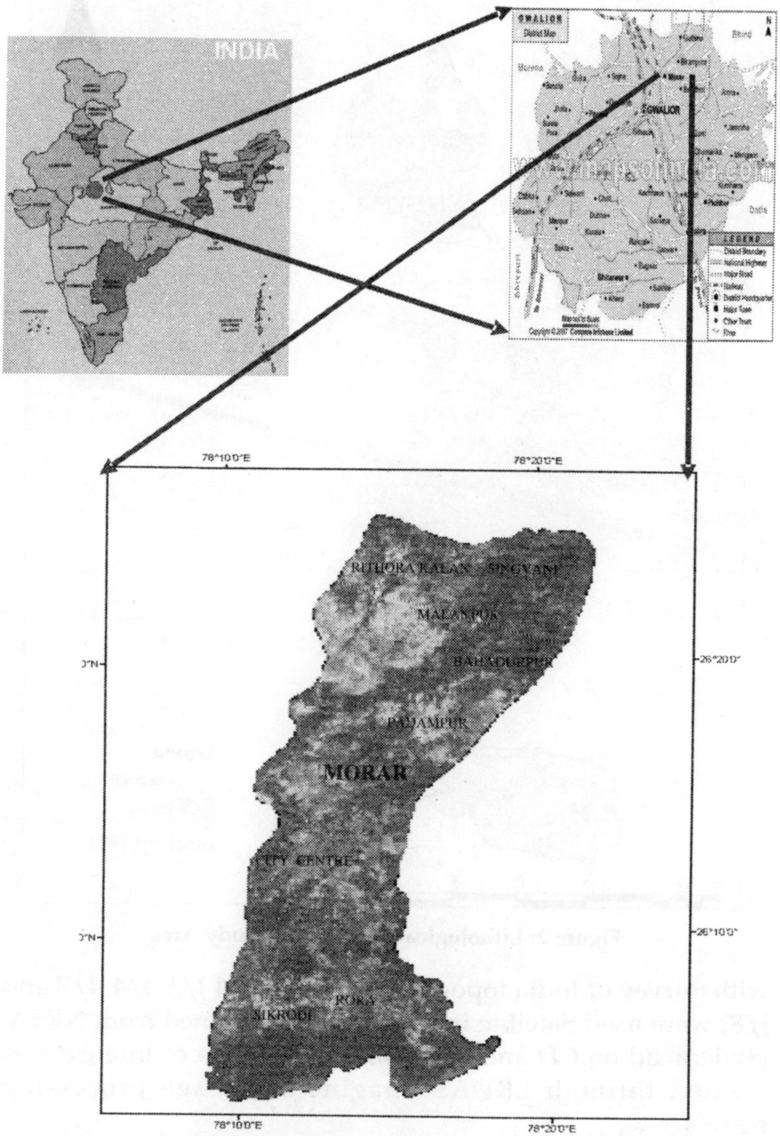

Figure 1: Location Map of the Morar Watershed, District Gwalior (M.P.), India

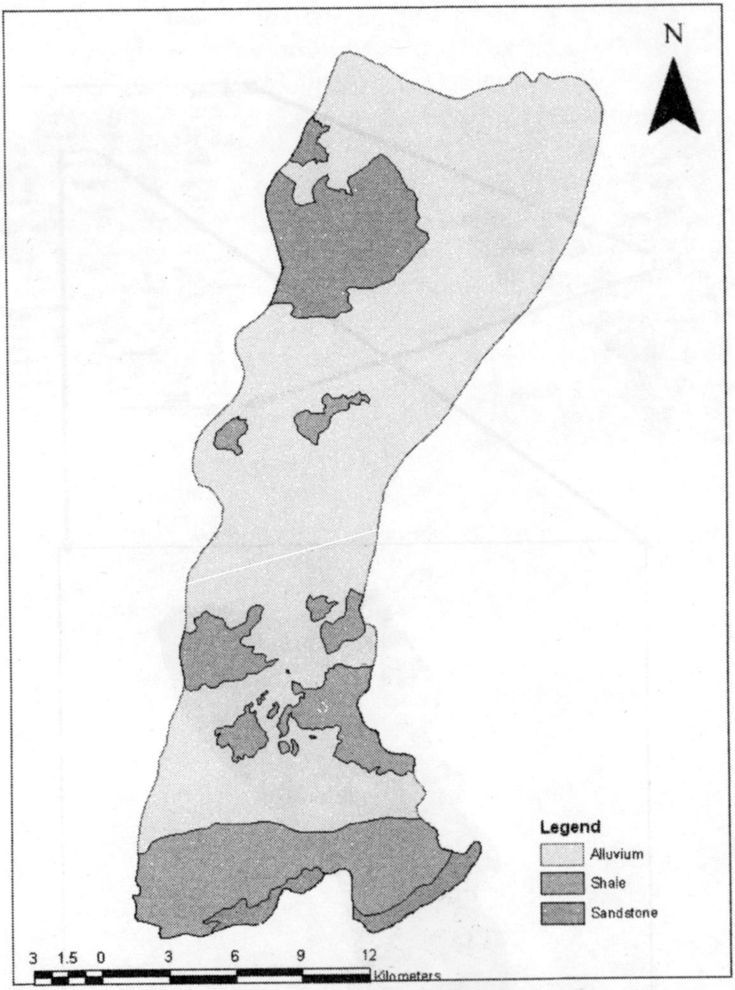

Figure 2: Lithological Map of the Study Area

with Survey of India topographical maps (54 J/3. J/4, J/7 and J/8) were used.Satellite imageries were obtained from NRSA, Hyderabad on CD and a sub set for the Area of Interest was created through ERDAS imagine 8.6 image processing software.

Geometric correction includes correction for geometric distortions due to sensor, earth geometry variations and conversion of data to real world coordinates on the Earth's surface. The IRS ID LISS-III and PAN digital data were geometrically rectified and registered with reference to SOI topographical maps on 1:5,0000 scale using Erdas Imagine 8.6 image processing software through the image to image registration techniques (Saraf, 1999). Various options of image enhancement techniques were tried to get the best image for visual interpretation. Histogram equalized stretch enhancement techniques were applied to the imagery of the study area for better interpretation of different features in the satellite imagery. All the computational work like digitization, editing, topology creation were done using Arc GIS (9.0) version software. Finally, all the land use/land cover of the area were on screen digitized and respective results were obtained (Figure 3) and (Table 1). Detailed methodology adopted in present paper is shown in the (Flow Chart 1).

Table 1
Area Covered Under different Land Use/Land Cover Categories

S.No	*Land use / Land cover*	*Area(in sq km)*	*Area (in %)*
1.	Urban Settlement	31.87	8.87
2.	Rural Settlement	5.82	1.44
3.	Industrial Complex	20.95	5.92
4.	Dense Forest	50.76	14.04
5.	Open Forest	11.74	2.90
6.	Forest Blank	1.85	2.11
7.	Agricultural Land	61.51	16.71
8.	Fallow Land	110.53	29.33
9.	Land with Scrub	31.25	7.73
10.	Land without Scrub	46.04	11.39
11.	Gullied/ Ravenous Land	5.05	1.49
12.	Reservoir	1.40	0.34
13.	River	6.08	1.50
14.	Ponds	1.62	0.64

RESULTS AND DISCUSSION

Land is one of the most important natural resource. All agricultural, animal and forestry production depend on the

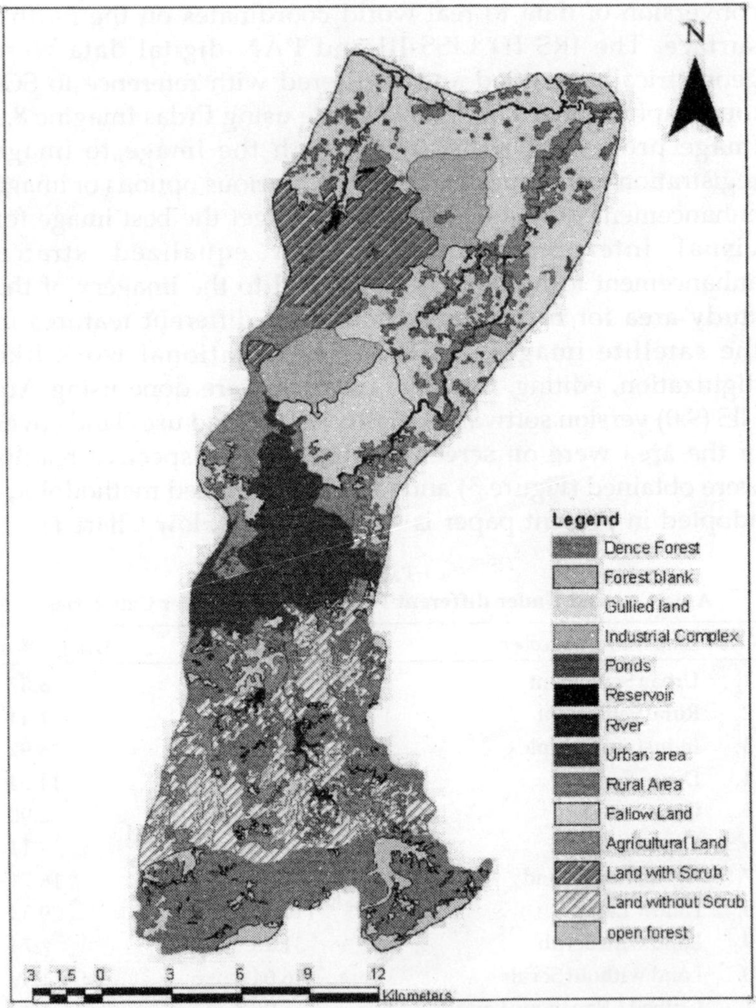

Figure 3: Land Use/Land Cover Map of the Study Area

productivity of the land. The landuse pattern and its spatial distribution are the prime requisites for the preparation of an effective landuse policy needed for the proper planning and management of any area. Various landuse categories of Morar watershed and their spatial distribution are given below (1.1) and Figure (3).

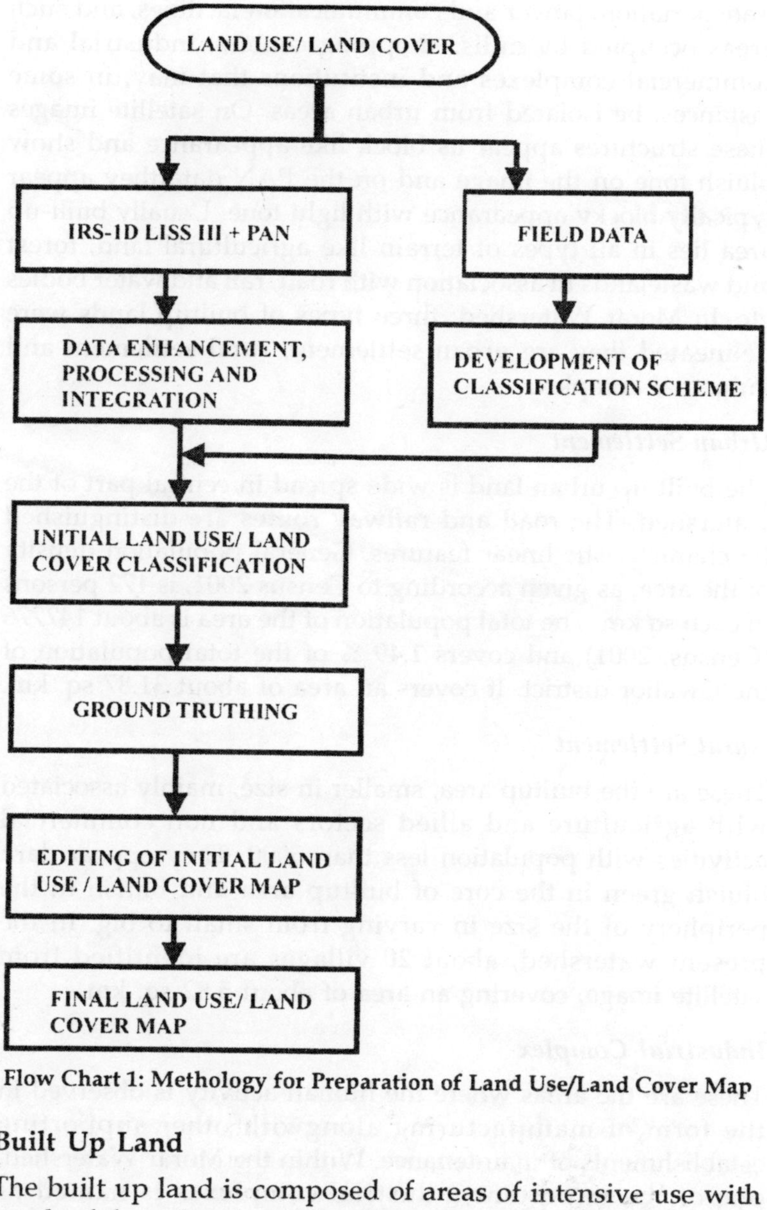

Flow Chart 1: Methology for Preparation of Land Use/Land Cover Map

Built Up Land

The built up land is composed of areas of intensive use with much of the urban/rural land use categories including cities, towns, villages, strip developments along highways, transportation, power and communication facilities, and such areas occupied by mills, shopping centers, industrial and commercial complexes and institutions that may, in some instances, be isolated from urban areas. On satellite images these structures appear as block like appearance and show bluish tone on the image and on the PAN data they appear typically blocky appearance with light tone. Usually built-up area lies in all types of terrain like agricultural land, forest and wastelands in association with road, rail and water bodies etc. In Morar Watershed, three types of builtup lands were delineated they are urban settlement, rural settlement and industrial complex.

Urban Settlement

The built-up urban land is wide spread in central part of the watershed. The road and railway routes are distinguished by characteristic linear features. General population density of the area, as given according to Census 2001, is 172 persons in each sq km. The total population of the area is about 147776 (Census, 2001) and covers 1.49 % of the total population of the Gwalior district. It covers an area of about 31.87 sq. km.

Rural Settlement

These are the builtup area, smaller in size, mainly associated with agriculture and allied sectors and non commercial activities with population less than 5000. They appear dark bluish green in the core of builtup area and bluish in the periphery of the size in varying from small to big. In the present watershed, about 20 villages are identified from satellite image, covering an area of about 5.82 sq. km.

Industrial Complex

These are the areas where the human activity is observed in the form of manufacturing alongwith other supporting establishments of maintenance. Within the Morar Watershed, one of the important industrial complexes is named as Malanpur Industrial Complex covering an area of about 20.95 sq. km.

Forest

Forest comprises of thick and dense canopy of tall tress, which predominantly remain green throughout the year. These lands are discerned by their red to dark red tone and varying in sizes. They show irregular shape and smooth texture. These forest areas are found on lower side of the watershed. Based on the tonal and textual variations, the forests of the watershed are divided into three categories as dense, open and forest blank.

Dense Forest

These categories of forest area cover canopy/density more than 40%. These forests have a characteristic dark red tone, smooth texture, and are in irregular shape. In the study area, such dense forests are found in lower and central side of the area at upland topography. Approximately such dense forests cover an area about of 50.76 sq. km.

Open Forest

This category of forest area covers canopy/density range between 10-40 % and the concentration of trees is less. They have the dark red to medium red tone, irregular shape and medium texture. Such open forests cover an area of about 11.74 sq. km

Forest Blank

These are the openings within forest areas, devoid of tree cover, and observed as openings of assorted size and shape as manifested on the imagery. They appear in light yellow to light brown tone generally small in size and most of these forests are found on hill tops and slopes. In the present study they account very small in size having an area of about 1.85 sq.km.

Agricultural Land

Agricultural land may be defined broadly as land used primarily for production of food and fiber. On high altitude imagery, the chief indications of agricultural activity will be distinctive by geometric field and road patterns on the landscape. By the help of satellite image it is possible to identify various agricultural lands up to level III. The information about kharif, rabi and double cropped areas were identified with the help of multidate satellite data. They show pink color with smooth appearance on pan data and dark patches with step like arrangement and agricultural land without crops show bluish/greenish grey with smooth texture. These can be easily separated from surrounding forest areas because of their regular shape. In the present investigation only kharif crops were identified. In Morar watershed such cropped areas were found mostly in the northeastern portion and some amount in southwestern portion. Such cropped areas cover approximately 61.51 sq.km.

Fallow Land

These are the lands, which are taken up for cultivation but are temporarily allowed to rest, un-cropped for one or more seasons, but not less than one year. These were identified by there dark greenish tone, smaller size, regular shape and medium texture. Such fallow lands are to be found well distributed in the central and northeastern portion of the watershed which occupy 110.53 sq.km.

Waste Land

It is described as degraded land which can be brought under vegetation cover with reasonable efforts and which is currently underutilized as well as the land which is deteriorating for lack of appropriate water and soil management or natural causes. Eroded soils, ravines, rocks, sand dunes and active depositional features are included in this category. In the study area three categories of wasteland are identified from satellite image and they cover approximately 83.34 sq.km.

Land with Scrub

These include the uplands or high grounds with scrub. These lands are subjected to degradation or erosion and consist mainly of thorny bushes. Such areas were identified from their yellowish tone, association with uplands and irregular shapes. These areas are found all over the watershed and the total area under this category is about 31.25 sq.km.

Land without Scrub

These lands are also found associated with higher topography and are formed by degradation or erosion. It could be identified from the satellite image from its light yellow tone, and its association with higher altitude. The absence of vegetation distinguishes this category from the earlier one. In the lower part of the watershed such lands are found to occur and cover about 46.04 sq.km.

Gullied/Ravenous Land

Gullies are formed as a result of localized surface run-off affecting the unconsolidated material resulting in the formation of perceptible channels causing undulating terrain. They are commonly found in sloping lands, developed as a result of concentrated run-off. They appear light yellow to bluish green in colour depending on the surface moisture and depth of erosion. They vary in size, shape with irregular broken network pattern. They are mostly associated with stream courses and sloping grounds with good rainfall and entrenched drainage. In the present study, these lands are covering an area of about 5.05 sq.km.

Water Body

Both man made and natural features covered with water are included in this category, i.e. rivers/streams, lakes, tanks, ponds and reservoirs. These water features appear black in colour tone on the satellite imagery and shallow water and deep water appear in light blue to dark blue in colour depending upon the depth of water body. In the study area, a number of ponds and one reservoir named Ramaua with one perennial and one seasonal river are delineated from the satellite image. The area covered under water body of the watershed is 9.10 sq.km.

CONCLUSIONS

Remote sensing techniques, now a days have the unique capability to map the land use/land cover quickly and accurately. The study has demonstrated the usefulness/ application potential of remote sensing techniques in assessing land use/land cover of the area. Agricultural area is mostly found in the northern portion of the watershed. It is observed that most of the southern portion of the watershed occupied by the forest cover over the denudational hills. These are the major portion of the watershed affected by severe soil erosion due to the occurrence of Shales and having steep slope. There is urgent need to manage soil erosion applying various techniques of soil conservation and large-scale aforestation. For ensuring planned development and monitoring the land utilization pattern, preparation of land use/land cover maps are necessary in the area.

ACKNOWLEDGEMENT

The authors are thankful to Head, School of Studies in Earth Science, Jiwaji University, Gwalior for providing necessary facilities for the present investigation. The authors would like to thank Dr. R. R. Navalgund, Director, Space Application Centre (SAC) Ahmedabad for granting permission to carry out Image Processing Work at SAC and to DR. A. S. Rajawat, Scientist SG MWRG/ RESIPA for his help and comments.

The first author expresses his gratefulness to the Vice Chancellor, Jiwaji University , Gwalior for granting University Research Fellowship for this purpose. The authors also acknowledge the unknown reviewers for their critical comments and suggestions.

REFERENCES

Balak Ram., and Kolarkar A. S. (1993), Remote Sensing Application in Monitoring Land Use Changes in Arid Rajasthan, *International Journal of Remote Sensing*, 14(17): 3191-3200.

Balaselvakumar S., Kumaraswamy K., Srileka S., and Raj Jawahar N. (2003), Remote Sensing Techniques for Land Use Mapping of Arjuna Basin, Tamilnadu, Map Asia Conference.

Chaudhary B. S., Saroha G. P., and Yadav Manoj (2008), Human Induced Land Use/Land Cover Changes in Northern Part of Gurgaon District, Haryana, India: Natural Resource Census Concept, *J. of Human Ecol.*, 23(3): 243-252.

Chaurasia, R., Loshali, D. C., Dhaliwal, S. S., Minakshi, Sharma, P. K. Kudrat, M. and Tiwari, A. K. (1996), Land Use Change Analysis for Agriculture Management - A Case Study of Tehshil Talwandi Sabo, Punjab, Photonirvachak, *Journal of the Indian Society of Remote Sensing*, 24(2): 115-123.

National Remote Sensing Agency (1989), Manual of Nation Wide Land Use/Land Cover Mapping Using Satellite Imagery, part-I, NRSA, Hyderabad.

Pateria Brijendra, and U. C. Singh (1994), Geology of Gwalior Area, Asian *J. Exp. Sci.*, 8(1&2): 9-13.

Pathak R. C., and J. M. Kate (1988), Land Use Mapping by Air Photo Techniques, *Journal of Indian Society of Remote Sensing*, 16(20): 47-52.

Rao D. P. (1991), IRS-IA Application for Land Use/Land Cover Mapping in India, *Current Science*, 69(3&4): 153-167.

Saha A. K., Arora M. K., Csaplovics E., and Gupta R. P. (2005), Land Use Cover Classification Using IRS LISS-III Image and DEM in a Rugged Terrain: A Case Study in Himalayas, *Geocarto International*, 20(2): 33-40.

Shrivastava S. K., and Gupta R. D. (2003), Monitoring of Changes in Land Use/Land Cover Using Multi - Sensor Satellite Data, Map India Conference.

19

TRENDS IN AGRICULTURE DURING PRE AND POST REFORM PERIOD IN INDIA

M. Ganesan

INTRODUCTION

Agriculture continuous to be the most crucial sector of Indian economy. It contributes about 26.8 per cent to the Gross Domestic Product (at current prices) and providing employment to nearly 2/3rd of the work force the new economic policy regime. It suffices to note that during the last 50 years, the country has achieved and sustained historically unprecedented growth of output (averaging somewhat 3 per cent per year against less than 1 per cent per annum during the first half of the century). Output growth has been well ahead of population. That this happened despite reaching the limit of expansion of cultivated land is significant. Expansion of cultivated area has virtually come to a stop in recent years. Growth in yields per hectare of cultivated area has become increasingly important, and is now the predominant source of growth. The virtual elimination of food grain imports can also be seen as an achievement.

In 1991, the government introduced a reform package to revitalize the economy. The economic reform measures have two objectives-namely short term measures and long-term measures. The short-term measures aimed at stabilization of the economy while the long term measures are designed for structural adjustment and reform, which is called as liberalization, privatization and globalization of economy. The present paper made an attempt to study the growth of agriculture in India during pre and post reforms periods.

A decade after the inception of economic liberalization, instead of experiencing an unprecedented boom in growth of the agricultural sector. India is facing some serious crisis. One of the very distributing features of growth pattern in the post liberalization period is very low employment generating potential of growth in secondary and territory sectors. The result is increase in concentration of workforce in agriculture and a persistent deterioration of relative productivity and income of workers engaged in the territory sector. The most significant limitation of liberalization policies followed so far is that it has failed to become an engine of rapid growth of agriculture and generate more employment. It has failed to make any visible dent in rural poverty or on unemployment.

In keeping view, the deceleration of growth rate in agriculture during 90's exists a strong case for increasing investment in rural infrastructure. It is essential to accord very high priority to public sector investment in agriculture which has been neglected during the last two decades. Since, there exists a strong complementation between public and private sector investment in agriculture; increasing public sector investment is also likely to cherish private investment. The availability of inadequate institutional credit is another means that would cherish investment especially by the small and marginal farmers. Streamlining the procedures, cutting down the red tape and involvement of village panchayats would help local participation. India being a founder member of the WTO is bound to undertake further economic reforms in agriculture. These would include removal of barriers to internal trade in agricultural commodities, abolition of zonal restrictions and compulsory procurement, opening future markets and protecting patent rights etc. But India should also do hard bargaining on the issue of market access and removal of subsidies by the developed countries. But the real gains from trade can only accrue if the country improves its competitiveness by increased productivity. This requires large investment in rural infrastructure, agricultural research and biotechnology, and creation of institutional arrangements of reaching benefits of research to all the cultivators including the small farmers.

However, the agriculture forms the back bone of Indian Economy and despite concerted industrialization in the last five decades; agriculture occupies a place of pride. It not only self sufficient in grain production but also has a substantial reserve. The Agricultural Development strategy was revised in 1999, as the national strategy on Sustainable Agriculture and Rural Development (SARD). The strategy is essentially based on the policy on food security and alleviation of hunger. The challenge of meeting the needs of present generations without compromising the ability of future generations to meet their needs was moulded the commission into the concept of sustainable development. In connection with the growth of agriculture, there is a need to analyse in two ways, such as (i) to analyze the reforms in the agriculture sector and the important step towards increasing growth rates in the Indian Economy and thus reducing poverty sustainability; and (ii) to analyze the trend of agricultural production in respect of pre and post reforms periods in India and offer suitable remedies to increase agricultural production.

Methodology

The New Economic Policy Liberalization policies were adopted in a big way in 1991, yet the reforms period (1980-81 to 2004-05) has been taken, so as to provide a fairly long time period for estimating the statistical and economical functions. This study is based on secondary data. Secondary data were collected from various issues. Growth rate, regression value and coefficient of variation have been employed in order to strengthen the findings of the study.

TRENDS IN AGRICULTURAL PRODUCTION

Agriculture has been mainstay of the Indian economy for centuries till the 1990's. The share of agriculture declined to

Table 1
Trends in Area, Production and Productivity of Agriculture During Pre and Post Reform Period

Year	Production(in 000 tonnes)	Area (million hectares)	Yield (kg / hectare)
	Pre-reform period		
1980-81	129.59	126.67	1023.00
1981-82	133.50	129.14	1032.00
1982-83	129.52	125.10	1035.00
1983-84	152.37	131.16	1162.00
1984-85	145.54	126.67	1149.00
1985-86	150.44	128.02	1175.00
1986-87	143.22	127.20	1128.00
1987-88	140.35	119.69	1173.00
1988-89	169.92	127.67	1331.00
1989-90	171.04	126.77	1349.00
1990-91	176.39	127.84	1380.00
AVG	149.26	126.90	1176.09
R Square	0.99		
CV	11.21	2.24	10.85
	Post-reform period		
1991-92	168.38	121.87	1382.00
1992-93	179.48	123.15	1457.00
1993-94	184.26	122.75	1501.00
1994-95	191.50	123.86	1546.00
1995-96	180.42	121.01	1491.00
1996-97	199.44	123.58	1614.00
1997-98	192.26	123.85	1552.00
1998-99	203.61	125.17	1627.00
1999-00	209.80	123.10	1704.00
2000-01	196.81	121.05	1626.00
2001-02	212.85	122.78	1734.00
2002-03	174.77	113.86	1535.00
2003-04	213.19	123.45	1727.00
2004-05	198.36	120.00	1652.00
2005-06	208.60	121.60	1715.00
2006-07	211.78	124.07	1707.00
AVG	195.34	122.19	1598.12
R Square	0.99		
CV	7.41	2.13	6.72

Source: RBI, Bulletin, 2007

59 per cent in 1950-51 to national income. Agricultural contribution of India as percentage of GDP (Gross Domestic Product) is quit high 26.8 per cent in relation to other develop countries like U.S.A (4%), Japan (3%) and France and UK (2%). It is after argued that the agricultural production increases accordance with the productivity. To test the implication of the inputs of the agriculture, of the productivity and the area under cultivation, a simple descriptive analysis was also attempted in this paper for both the time period. Table 1 presents the estimated co-efficient of the test for both pre and post reform periods. The obtained significant co-efficient of production 11.2 per cent and 7.4 per cent in the pre and post reform period, respectively, post that the production department significantly on the productivity. Therefore, it can be concluded that productivity causes production particularly in the post reform period. Moreover, the obtained significant co-efficient of the area under production reveals 2.24 per cent and 2.13 pre cent during pre and post reform period respectively. In regards to yield, obtained significant value is that 10.85 per cent and 6.72 per cent during the pre and post reform period respectively. It is evident that the production of agricultural produce is mainly due to the increase of productivity. The land under cultivation is decreased over the period, than the pre reform period.

TRENDS IN NET AVAILABILITY OF FOOD GRAINS

The data reveals that the annual net availability of the food grains were 150 kgs. In 1980, it increased to 165 kgs. in 1985 it reached the peak of 180 kgs. in the year 1989. It declined thereafter to 154 kgs. in 2005. Table 2 presents the growth of food grains during pre-reform period (1991-92 to 2004-05). In absolute terms, the total food grains increased form 150 kg per capita per annum in 1980, there by showing roughly a growth of 1.2 times between these time periods. The growth of food grains showed fluctuating trends during the pre reform period. It was found varying from a positive growth of 9.72 per cent in 1984 to a negative growth of –4.22 per cent in 1990. But, the estimated negative growth rate particularly during pre-reform period could be attributed to various reasons. These concluded that the depression in Indian economy had emerged as a major challenge to the Indian agriculture. The food crisis during the 1960s that virtually caused to the import of food grains is 21.9 lakh tones was major challenge to the Indian economy. During the pre-reform period, an absolute increase in food grains was estimated with an average annual increase of 1.27 kgs. per capita per annum. The trend growth rate estimated for the food grains was 167 kgs. per capita per annum.

The intensive implementation of the reform process after 1991 made India to experience a radical shift in the structure of agricultural production. The foreign countries structure to invest more in Indian industries as the reform measures turned unfavoured to the Indian agriculture. The kgs. per capita per annum slightly increased in the first year of the introduction of reforms in 1991 (186 kg. per annum) as compared to the previous years 150 kgs per capita per annum. The food grains availability was more than 180 kgs. per capita per annum except in 1991 and 1997 during which food grains registered negative growth at –7.94 per cent and –5.82 per cent respectively. The food grains per capita per annum is lower. Various recent studies on Indian Agriculture reveals that the current level of agricultural contribution to the GDP was too less, when compared to the pre-reform period, due to the implementation of the WTO agreement on agriculture. The obtained significant co-efficient of the food grains availability is 29 per cent and 25 per cent in pre and post reform period respectively.

The implementation of atomic approval system, raising the limit of the FDI and expansion of areas for MNCs and TNCs participation as a part of reform process, tremendously affected the agricultural growth in to the Indian economy. During pre-reform period, the average foodgrains was estimated 167 kgs. per capita per annum, where as the same was estimated to be 159 kgs per capita per annum in the post reform period. The reduction of foodgrains production was proved by estimating the co-efficient of the variance, it was

25 per cent less in the pre reform period evident that decreasing of foodgrains in India.

SUGGESTIONS

Since the reforms in agricultural sector are likely to open up export possibilities, there is need to design a proactive policy to cherish exports and involve the small and marginal farmers in increasing productivity and in deriving benefits from increased agricultural exports. There is a need to create innovative institutions including integrated cooperatives like the mother dairy and other service cooperatives. Special efforts should also be made to develop new technologies for the farming sector and reach these to the small farmers for enabling them to diversify their production towards high value commercial and export commodities. The efforts on the production front should be supplemented by creation of institutions like trading houses; market intelligence services and creation of net work of information on national and international prices, reforms in agricultural sector are likely to open up export possibilities. There is also a need to create necessary infrastructure in processing marketing and grading produce. Investment in information and infrastructure through market committees would percolate the information to the local levels.

CONCLUSION

The changes in the Indian economic policies in general and agreement of agriculture significantly to lead the poverty in the country. The estimated co-efficient of the production of agricultural product values emphasizes the dependence of economic growth on the levels of agricultural production. The changes in the sectoral development in India reflect the success of a selective regime. As a result, the average annual growth recorded in area under cultivation during the post-reform period 1991-2005 was 122.1 compared to 126.9 in 1980-1990. It is mainly due to the implementation of the WTO Agreement on Agriculture. It can be faced adopting the above said suggestions through revamping the second phase of Green Revolution in India.

Table 2
Trends in Area, Production and Productivity in Indian Agriculture during Pre- and Post-Reform Period

Year	Rice	Annual Growth (%)	Wheat	Annual Growth (%)	Other Cereals	Annual Growth (%)	Cereals	Annual Growth (%)	Gram	Annual Growth (%)	Pulses	Annual Growth (%)	Per capita Food grains	Annual Growth (%)
1980	60.8	-	46.4	-	31.7	-	138.9	-	3.9	-	11.3	-	150.2	-
1981	72.2	18.75	47.3	1.94	32.8	3.47	152.3	9.65	4.9	25.64	13.7	21.24	166	10.52
1982	70.5	-2.35	46.7	-1.27	34.6	5.49	151.8	-0.33	5.1	4.08	14.3	4.38	166.1	0.06
1983	62	-12.06	52.7	12.85	30.4	-12.14	145.1	-4.41	5.7	11.76	14.1	-1.40	159.5	-3.97
1984	72.2	16.45	51.4	-2.47	36.1	18.75	159.7	10.06	5	-12.28	15.3	8.51	175	9.72
1985	68.9	-4.57	50.6	-1.56	32.1	-11.08	151.6	-5.07	4.7	-6.00	13.9	-9.15	165.5	-5.43
1986	77.4	12.34	55.1	8.89	25.8	-19.63	158.3	4.42	5.9	25.53	16	15.11	174.3	5.32
1987	75.2	-2.84	57.6	4.54	25.9	0.39	158.7	0.25	4.5	-23.73	13.3	-16.88	172	-1.32
1988	68.7	-8.64	56.3	-2.26	25.1	-3.09	150.1	-5.42	3.5	-22.22	13.3	0.00	163.4	-5.00
1989	78.5	14.26	57.0	1.24	29.3	16.73	164.8	9.79	4.9	40.00	15.3	15.04	180.1	10.22
1990	77.4	-1.40	48.4	-15.09	31.7	8.19	157.5	-4.43	3.9	-20.41	15	-1.96	172.5	-4.22
CGAR (%)	2.22		0.38		0.00		1.15		0.00		2.61		1.27	
AVG	71.25		98.90		30.50		153.53		4.73		14.14		167.69	
CV	29.93		29.91		31.03		29.22		32.53		30.10		29.25	
1991	80.9	4.52	60	23.97	29.2	-7.89	171	8.57	4.9	25.64	15.2	1.33	186.2	7.94
1992	79.2	-2.10	57.9	-3.50	21.5	-26.37	158.6	-7.25	3.7	-24.49	12.5	-17.76	171.2	-8.06
1993	73.4	-7.32	51.2	-11.57	31.6	46.98	156.2	-1.51	3.9	5.41	13.2	5.60	169.4	-1.05
1994	75.7	3.13	58.2	13.67	24.5	-22.47	158.4	1.41	4.3	10.26	13.6	3.03	172	1.53
1995	80.3	6.08	63	8.25	23.7	-3.27	167	5.43	5.4	25.58	13.8	1.47	180.8	5.12
1996	74.6	-7.10	64.3	2.06	22.6	-4.64	161.5	-3.29	4.1	-24.07	12	-13.04	173.5	-4.04
1997	78.1	4.69	65.4	1.71	26.6	17.70	170.1	5.33	4.5	9.76	13.3	10.83	183.6	5.82

Contd.

Year	Rice	Annual Growth (%)	Wheat	Annual Growth (%)	Other Cereals	Annual Growth (%)	Cereals	Annual Growth (%)	Gram	Annual Growth (%)	Pulses	Annual Growth (%)	Per capita Food grains	Annual Growth (%)
1998	73.1	-6.40	55.3	-15.44	22.8	-14.29	151.2	-11.11	4.9	8.89	11.6	-12.78	163.2	-11.11
1999	74.2	1.50	59.2	7.05	23.1	1.32	156.7	3.64	5.3	8.16	10.9	-6.03	170	4.17
2000	74.3	0.13	58.4	-1.35	21.5	-6.93	154.3	-1.53	3.9	-26.42	12.9	18.35	165.9	-2.41
2001	69.5	-6.46	49.6	-15.07	20.5	-4.65	141	-8.62	2.9	-25.64	10.6	-17.83	151.9	-8.44
2002	83.5	20.14	60.8	22.58	23.1	12.68	167.4	18.72	3.9	34.48	12.9	21.70	180.4	18.76
2003	66.2	-20.72	65.8	8.22	17.1	-25.97	149.1	-10.93	3.1	-20.51	10.6	-17.83	159.7	-11.47
2004	71.3	7.70	59.2	-10.03	25.3	47.95	155.8	4.49	4.1	32.26	13.1	23.58	168.9	5.76
2005	64.7	-9.26	56.3	-4.90	21.7	-14.23	142.7	-8.41	3.9	-4.88	11.5	-12.21	154.2	-8.70
CGAR	-1.48		-0.42		-1.96		-1.20		-1.51		-1.84		-1.25	
AVG	69.94		55.29		22.17		147.56		3.92		11.73		159.43	
CV	25.93		26.16		28.88		25.62		30.09		26.99		25.66	

Source: RBI, Bulletin, 2007.

REFERENCES

Agrawal A. N. (2000), Indian Economy, *Problems of Development and Planning*, New Delhi: Wiley Eastern Ltd.

Ahluwalia I. J., and I. M. D. Little (2006), India's Economic Reforms and Development Essays for Manmohan Singh, New Delhi: Oxford University Press.

Bansil P. C. (1981), Agricultural Problems of India, New Delhi: Oxford and IBH publishing Co.

Benjamin R. E., Hariharan, and Karunagaran M. (1989), Economics of Agriculture, New Delhi: S.Chand & Co.

Ishwar C. Dhingra (2006), The Indian Economy Environment and Policy, New Delhi: Sultan Chand & Sons.

Mamoria C. B. (1999), Agricultural Problems of India Allahabad: Kitab Matal.

Misra, and Puri (2004), A Text Book of Indian Economy, Delhi: Himalaya Publishing House.

Raj Kapila, and Uma Kapila (2002), India's Economy in the 21st Century - A Collection of Select Articles, New Delhi: Academic Foundation.

Ruddar Datt, and Sundharam K. P. M. (2007), Indian Economy, New Delhi: S. Chand & Co.

Sankaran S. (2002), Indian Economy Problems, *Policies and Development*, Chennai: Margham.

Tyagi B. P. (2000), Agricultural Economics and Rural Development, Meerut: Jai Prakash Nath & Co.,

20

REGIONAL VARIATION IN RURAL DEVELOPMENT IN CUDDALORE DISTRICT

T. Sezhiyan

INTRODUCTION

As regards measuring development, most of the scholars used the percapita income and/or consumption expenditure as a proxy indicator for level of living. The percapita income which is widely accepted as a general measure of development, is an index of production not consumption. Society can be made better off by any number of changes–length of life or mortality rates - even without an increase in GNP. Kuznets explicitly notes that "from the standpoint of basic wants or needs, the system of valuation in national income measures may lead to an underestimate of the extent of economic progress or of the extent of economic superiority across space in satisfying primary wants". Further, income as an aggregation concept does not reflect society's physical qualities of life, such as life expectancy, its birth, death and morbidity characteristics, and its literacy. So income is not a very satisfactory measure of development and at best it can be taken as one of the indicators of development. On the other hand, to assess the poverty level and also development of the particular sector, a few authors and governmental organisations have used food consumption expenditure as a proxy indicator of development.

The problems of using consumption expenditure are: (1) the estimate of consumption at different levels are taken into account only in private consumption. But no one includes what an individual or household shares socially financed consumption. (2) It has been pointed out that consumptions as an indicator of level of living can be misleading as it will tend to underestimate the differences between those at low and high levels. This paper analyses regional variation in rural development in Cuddalore district on the basis of chosen indicators.

Method of Analysis

This district is divided in to three regions viz. Flood Plain Region, Coastal Region and North Plain Region. Flood Plain region has five block viz. Kumaratchi block, Kattumannar koil block, Keerapalayam block, Melbhuvanagiri block and Kammapuram block. Coastal region has 3 blocks viz. Kurinjipadi block, Parangipettai block and Cuddalore block, North Plain region has 5 blocks viz. Mangalore block, Vridhachalam block, Nallur block, Annagramam block and Panruti block. This study confines to six selected blocks. In this study two blocks from each region have been selected for the purpose of analysis. The selected include, Bhuvanagiri, Kattumannar Koil, Kurinjipadi, Cuddalore, Mangalore and Panruti blocks. It could be noted that from each block two village are selected and thus totally 6 villages are selected from six blocks. Atarmangalam village is selected from Kattumannar Koil block and Erumpur village is selected from Bhuvanagiri block. Sandorpalayam villages is chosen from Cuddalore block and Ayyekuppam vilalge is chosen from Kurinjipadi block. Passer Village is chosen from Mangalore block and L. Puram village is selected from Panruti block. The development of rural infra structure has been analysed in the three dimension in Cuddalore district on the basis of chosen indicators.

RESULTS AND DISCUSSION

A study of data in Table 1 indicates the region wise development status in Cuddalore district.

Table 1 presents data on the region wise development status in Cuddalore district. The population density of Cuddalore district is 657.83 persons. Among the three region in Cuddalore district Coastal region has highest population density and flood plain region has lowest population density. The percentage of urban population to total population is 5 per cent. The flood plain region and north plain region have less per cent of urban population to the total population in the district. This district has 977.67 females per 1000 males. The coastal region has lowest sex ratio. The percentage of scheduled caste total population is 13.26 per cent in Cuddalore district. Out of the total three regions in Cuddalore district, north plain region has lowest per cent of scheduled caste population. This district has lowest percentage of scheduled tribe population and all regions have below 1 per cent of scheduled tribe population. In this district , average size of household is 4.25. The flood plain region has average household size of 4 members. The percentage of workers to total population in Cuddalore district is worked out to 22.73 per cent. The coastal region occupies the first position with respect to percent of workers to total population. In Cuddalore district the percentage of female workers to total workers is worked out to 8.04 per cent. The flood plain region has lowest per cent of female workers to total workers. It could be noted that Cuddalore district has 23.63 per cent of agriculture workers to the total workers. The percentage of agriculture workers to the total workers is lowest in flood plain region and highest in coastal region. In Cuddalore district the percent of agriculture labourers to agriculturalists is worked out to 23.34 per cent. The percent of agriculture labourers to agriculturalists is worked out to 12.40 per cent in flood plain region and it is 38.26 per cent in coastal region. Cuddalore district has 6.36 per cent of workers in manufacturing industry with respect to total number of workers.

Table 1
Region Wise Development Status in Cuddalore District

S.No.	Socio – Economic indicators	Flood Plain Region	Coastal Region	North Plain	District Average
1.	Density of population	520.50	757.00	696.00	657.83
2.	Percentage of Urban population to total population	10.27	16.62	11.5	12.80
3.	No. of females per 1000 males	981.00	959.00	993.00	977.67
4.	Percentage of schedule caste to total population	12.33	17.16	10.29	13.26
5.	Percentage of schedule tribe to total population	0.21	0.38	0.29	0.29
6.	Average size of household	4.00	4.50	4.25	4.25
7.	Percentage of workers to total population	21.46	26.27	20.45	22.73
8.	Percentage of female workers to total workers	5.16	8.69	10.28	8.04
9.	Percentage agriculture workers to total workers	15.14	37.11	18.64	23.63
10.	Percentage of Agricultural labours to total Agricultural lists	12.40	38.26	19.35	23.34
11.	Percentage of workers in manufacturing industry to total workers	8.04	6.02	5.01	6.36
12.	Percentage of area sown to total area	21.74	32.88	18.61	24.41
13.	Percentage of gross cropped area to net area sown	11.66	30.88	16.07	19.54
14.	Per capita net area sown	16.09	19.50	11.20	15.60
15.	Percentage of net area irrigated to net area sown	24.94	50.08	32.31	35.78
16.	Percentage of gross area irrigated to gross cropped area	18.66	32.25	22.29	24.40
17.	Bovine population per Sq. Km	352.00	168.00	192.00	237.33
18.	No. of draught animals for thousand hectares of net area sown	513.50	249.00	400.50	387.67
19.	Percentage of villages electrified	0.91	0.94	0.94	0.93
20.	No. of post offices per thousand of population	17.50	11.55	19.00	16.02
21.	No. of radio sets per thousand of population	63.83	33.50	43.50	46.94
22.	Percentage of literacy	26.58	35.22	29.61	30.47
23.	Percentage of literacy amongst schedule castes	57.62	65.2	61.22	61.35
24.	Percentage of villages covered by protected water supply	0.94	0.94	0.96	0.95
25.	No. of households per 100 residential occupied houses	94	92	99	95

Source: Computed

Regarding agriculture 24.41 per cent of area sown to the total area. The area sown to the total area is highest in coastal region and lowest in north plain region. The percent of Gross cropped area to net cropped area is worked out to 19.54 per cent in Cuddalore district. The percent of Gross cropped area to net cropped area is highest in coastal region and lowest in flood plain region. Cuddalore district has 15.60 per cent of per capita net sown area. The per capita net sown area is lowest in north plain region and highest in coastal region. The per cent of net irrigated area to net sown area is worked out to 35.78 per cent in Cuddalore district. The coastal region has 50.08 per cent of net irrigated area to net sown area. The per cent of gross irrigated area to gross cropped area is worked out to 24.40 per cent in Cuddalore district. The coastal region has 32.25 per cent of gross irrigated area to gross cropped area and it is lowest in flood plain region. Cuddalore district has 237.33 bovine population per sq. km. The density of bovine population is highest in flood plain region and lowest in coastal region. In Cuddalore district, 93 per cent of the villages are electrified. There is no significant regional variation with respect to per cent of villages electrified in Cuddalore district. In this district, 387.67 draught animals per thousand hectares of net sown area. The number of draught animals per thousand hectares of net sown area is highest in flood plain region and lowest in coastal region. Cuddalore district has 16.02 post offices per 1000 population. The number of post offices per 1000 population is lowest in coastal region and highest in north plain region. This district has 46.94 radio sets per 1000 population. The flood plain region has highest number of radio sets and coastal region has lowest number of radio sets. The literacy rate of Cuddalore district is worked out to 61.35 per cent. The flood plain region has lowest level of literacy. In this district, 95 per cent of villages have protected water supply. There is no significant variation among regions with respect to coverage of protected water supply. The number of the households per 100 residential occupied is worked out to 95.

A study of data in Table 2 indicates the block wise development status in Cuddalore district. It could be noted that the population density is highest in Cuddalore block and lowest in Mangalore block. The Cuddalore block occupies the first position with respect to per cent of urban population and Kattumannar koil block comes to second position. The present of urban population is lowest in Mangalore block. However, Mangalore block has highest sex ratio and sex ratio is lowest in Kurinjipadi block. The concentration of scheduled caste population is highest in Mangalore block and lowest in Panruti block. The scheduled tribe population is considerably higher in Panruti block than those of other blocks chosen in the study. Kurinjipadi block occupies the first position with respect to possession of more number of households than those of other blocks and this block has more per cent of workers to the population. Panruti block has lowest number of female workers to the total population. Bhuvangiri block occupies the first position with respect to per cent of agriculture workers to the total workers and also per cent of agriculture labourers to total agriculturalists. Cuddalore block takes the first position with respect to more number of industrial workers than those of other blocks. The total cropped area is highest in Kattumannar koil block and lowest in Mangalore block. The per cent of gross cropped to net cropped area is highest in Mangalore block and lowest in Bhuvanagiri block. The per capita net sown area is highest in Kattumannar Koil block and lowest in Kurinjipadi block. The per cent of net area irrigated to net area sown is highest in Cuddalore block and lowest in Mangalore block. The percent of gross area irrigated to gross cropped area is highest in Kurinjipadi block and lowest in Panruti block. Kattumannar Koil block occupies the first position with respect to possession of bovine population and draught animals. Cuddalore block occupies the first position with respect to per cent of villages electrified and Bhuvanagiri block takes the last position. Kurinjipadi block has less number of post offices in relation to its population size. Cuddalore block has more number of radio sets and Panruti block has less number of radio sets per

Table 2
Block Wise Development Status in Cuddalore District

S. No.	Socio – Economic indicators	Flood plain region		Coastal region		North plain	
		Bhuvanagiri	*Kattumannar Koil*	*Kurinjipadi*	*Cuddalore*	*Mangalur*	*Panruti*
1.	Density of population	526	515	508	1006	328	1064
2.	Percentage of urban population to total population	26	28.69	24.00	38	14	23
3.	No. of females per 1000 males	983	979	933	985	1020	966
4.	Percentage of schedule caste to total population	32.00	34.21	31.82	34	39	18.70
5.	Percentage of schedule tribe to total population	0.70	0.58	0.29	0.75	12	5
6.	Average size of household	4	4	5	4	4	4.5
7.	Percentage of workers to total population	43	48.51	53.57	52	52	37.10
8.	Percentage of female workers to total workers	14.00	17.50	37.65	17	45	10.10
9.	Percentage agriculture workers to total workers	72	29.56	22.38	74	67	61.50
10.	Percentage of Agricultural labours to total Agriculturalists	77	2	51.75	76	46	23.2
11.	Percentage of workers in manufacturing industry to total workers	4.00	3.3	3.64	12	1	1.60
12.	Percentage of area sown to total	65.22	82.98	76.48	65	48	73.90

Contd.

S. No.	Socio – Economic indicators	Flood plain region		Coastal region		North plain	
		Bhuvanagiri	Kattumannar Koil	Kurinjipadi	Cuddalore	Mangalur	Panruti
	area						
13.	Percentage of gross cropped area to net area sown	59	72.36	76	61	118	95
14.	Per capita net area sown	0.12	0.17	0.11	0.14	16	0.4
15.	Percentage of net area irrigated to net area sown	99.40	88.21	70.43	99.45	33	29
16.	Percentage of gross area irrigated to gross cropped area	61.75	71.18	75.00	63.75	31	27
17.	Bovine population per Sq. Km	169	535	336	85	90	80
18.	No. of draught animals for thousand hectares of net area sown	360	667	108	390	762	39
19.	Percentage of villages electrified	88	93	89	99	92	96
20.	No. of post offices per thousand of population	12	23	10	13	23	15
21.	No. of radio sets per thousand of population	61	66.66	57	67	40	35
22.	Percentage of literacy	63.00	52	43.23	70	57.5	64.8
23.	Percentage of literacy amongst schedule castes	22.63	24.69	20.21	23.05	10.15	20.7
24.	Percentage of villages covered by protected water supply	92	96	99	88	99	93
25.	No. of households per 100 residential occupied houses	96	92	95	89	99	99

Source: Computed

population than those of other blocks. Among the selected blocks Cuddalore block has highest level of literacy and Kattumannar koil block lowest level of literacy. The per cent of literacy among the scheduled caste is highest in Kattumannar Koil block and lowest in Mangalore block. Cuddalore block is pushed down to last position with respect to percent of villages covered by protected water supply.

A study on data in Table 3 indicates the villages wise development indicators in Cuddalore district. It could be noted that the population density is highest in L. N. Puram village and lowest in Ayekuppam village. The Passar village the first position with respect to percent of urban population and L. N. Puram village comes to second position. The present of urban population is lowest in Ayekuppam village. However, Ayekuppam village has highest sex ratio and sex ratio is lowest in Atarmangalam village. The concentration of scheduled caste population is highest in Ayekuppam village and lowest in Passar village. The scheduled tribe population is considerably higher in Ayekuppam village than those of other blocks chosen in the study. Erumpur village occupies the first position with respect to possession of more number of households than those of other blocks and this village has more per cent of workers to the population. Passar village has lowest number of female workers to the total population. L.N. Puram village occupies the first position with respect to per cent of agriculture workers to the total workers. Atarmangalam village occupies the first position with respect to per cent of agriculture labourers to total agriculturalists. L. N. Puram village takes the first position with respect to more number of industrial workers than those of other blocks. The total cropped area is highest in Sandorpalayam village and lowest in Ayekuppam village. The percent of gross cropped to net cropped area is highest in Ayekuppam vilalge and lowest in Atarmangalam village. The per capita net sown area is highest in Ayekuppam village and lowest in Passar village. The per cent of net area irrigated to net area sown is highest in L. N. Puram village and lowest in Ayekuppam village. The percent of gross area irrigated to gross cropped

Table 3

Village Wise Development Indicators in Cuddalore District

S.No.	Socio-Economic indicators	Atarmangalam	Erumpur	Sandorpalayam	Ayekuppam	Passar	L. N. Puram
1.	Density of population	326	304	412	272	894	996
2.	Percentage of Urban population to total population	21	18.00	20.19	10	42	23
3.	No. of females per 1000 males	782	813	851	922	872	885
4.	Percentage of schedule caste to total population	24.00	22.18	14.21	29	11.60	19
5.	Percentage of schedule tribe to total population	0.70	0.29	0.58	8	4	0.65
6.	Average size of household	3	4	3	3	3.1	3.8
7.	Percentage of workers to total population	33	43.57	38.51	41	32.16	49
8.	Percentage of female workers to total workers	10.00	31.25	11.28	34	9.90	11
9.	Percentage agriculture workers to total workers	68	18.32	19.56	57	51.18	69
10.	Percentage of Agricultural labours to total Agricultural workers	68	42.76	1	33	18.90	63
11.	Percentage of workers in manufacturing industry to total workers	3.00	2.23	1.75	1	1.50	9
12.	Percentage of area sown to total area	59.21	62.48	83.98	33	63.81	59
13.	Percentage of gross cropped area to net area sown	55	69	63.88	111	89	64

Contd.

S.No.	Socio-Economic indicators	Atarmangalam	Erumpur	Sandorpalayam	Ayekuppam	Passar	L. N. Puram
14.	Per capita net area sown	0.09	0.07	11.69	14	0.3	0.06
15.	Percentage of net area irrigated to net area sown	79.52	52.43	78.21	23	20	89.45
16.	Percentage of gross area irrigated to gross cropped area	52.88	63.00	69.15	26	21	59.78
17.	Bovine population per Sq. Km	153	142	438	70	69	99
18.	No. of draught animals for thousand hectares of net area sown	280	96	487	520	38	370
19.	Percentage of villages electrified	92	93	88	90	96	89
20.	No. of post offices per thousand of population	11	0.15	23	0.23	15	13
21.	No. of radio sets per thousand of population	59	33	63	9	2	49
22.	Percentage of literacy	61.00	41.26	29.23	47.52	59.20	69.23
23.	Percentage of literacy amongst schedule castes	20.15	10.15	26.15	35.07	31.67	24.65
24.	Percentage of villages covered by protected water supply	89	99	92	96	98	93
25.	No. of households per 100 residential occupied houses	93	95	88	99	99	98

Source: Computed

area is highest in Sandorpalayam village and lowest in Passar village. Sandorpalayam village occupies the first position with respect to possession of bovine population and Ayekuppam village occupies the first position with respect to possession of draught animals. Passar village occupies the first position with respect to per cent of villages electrified and Sandorpalayam village takes the last position. Sandorpalayam village has less number of post offices in relation to its population size. Atarmangalam village has more number of radio sets and Passar vilage has less number of radio sets per population than those of other blocks. Among the selected blocks L. N. Puram village has highest level of literacy and Sandorpalayam village lowest level of literacy. The per cent of literacy among the scheduled caste is highest in Ayekuppam village and lowest in Erumpur village. Atarmangalam village is pushed down to last position with respect to percent of villages covered by protected water supply. The percent of number of households per 100 residential is highest in Ayekuppam village and Passar village and lowest in Erumpur village.

A study on data in Table 4 indicates the village wise distribution of agriculture of Infrastructure. Atarmangalam village has 63 agriculture infrastructure facilities. These include 46 diesel engine, 10 electric motor pump sets, 6 length of canal and 5 irrigation tanks. Erumpur village has 59 agriculture infrastructure facilities. These include 25 diesel engine, 23 electric motor pump sets, 5 length of canal and 6 irrigation tanks. Sandorpalayam village has 74 agriculture infrastructure facilities. These include 18 diesel engine, 42 electric motor pump sets, 10 length of canal and 4 irrigation tanks. Ayekuppam village has 80 agriculture infrastructure facilities. These include 35 diesel engine, 33 electric motor pump sets, 9 length of canal and 3 irrigation tanks. Passar village has 73 agriculture infrastructure facilities. These include 40 diesel engine, 25 electric motor pump sets, 6 length of canal and 2 irrigation tanks. L. N. Puram village has 89 agriculture infrastructure facilities. These include 45 diesel engine, 36 electric motor pump sets, 5 length of canal and 3 irrigation tanks.

A study of data in Table 5 indicates the village wise distribution of power infrastructure facilities. It could be noted that Sandor palayam village occupies the first position with respect to possession of power infrastructure facilities. It has 278 power infrastructure facilities and among them 20 street lights, 180 electrified households, 28 houses with one light scheme, 8 sodium lamps and 10 electric pump sets. Ayekuppam village comes to the second position with respect to possession of power infrastructure facilities. It has 272 power infrastructure facilities and among them 9 street lights, 179 electrified households, 33 houses with one light scheme, 8 sodium lamps and 33 electric pump sets. L. N. Puram village comes to the third position with respect to possession of power infrastructure facilities. It has 269 power infrastructure facilities and among them 15 street lights, 169 electrified households, 40 houses with one light scheme, 9 sodium lamps and 36 electric pump sets. Erumpur village comes to the fourth position with respect to possession of power infrastructure facilities. It has 256 power infrastructure facilities and among them 15 street lights, 176 electrified households, 36 houses with one light scheme, 6 sodium lamps and 23 electric pump sets. Passar village comes to the fifth position with respect to possession of power infrastructure facilities. It has 250 power infrastructure facilities and among them 18 street lights, 175 electrified households, 26 houses with one light scheme, 6 sodium lamps and 25 electric pump sets. Atarmangalam village comes to the last position with respect to possession of power infrastructure facilities. It has 209 power infrastructure facilities and among them 10 street lights, 162 electrified households, 22 houses with one light scheme, 5 sodium lamps and 10 electric pump sets.

A study of data in Table 6 indicates the village distribution of road facilities. The total road length in Atarmangalam village is 4011.34 meters. Out of the total road length tar road length is 1222 meters, length of cement road is 1366.98 meters and length of mud road is 1422.36 meters. Erumpur village has 4186.47 meters total length of road and among them length of length tar road length is 1233 meters, length of cement

Table 4
Village Wise Distribution of Agriculture Infrastructure

Village wise	*Atarmangalam*	*Erumpur*	*Sandorpalayam*	*Ayekuppam*	*Passar*	*L. N. Puram*
No. of Pump sets	42	25	18	35	40	45
No. of Electrical Motor pump sets	10	23	42	33	25	36
Length of canal	6	5	10	9	6	5
No. of tanks and ponds	5	6	4	3	2	3
Total no. of Infrastructure	63	59	74	80	73	89

Source: Computed

Table 5
Village Wise Distribution of Power Infrastructure Facilities

Village wise	*Atarmangalam*	*Erumpur*	*Sandorpalayam*	*Ayekuppam*	*Passar*	*L. N. Puram*
No. of street lights	10	15	20	19	18	15
No. of electrified households	162	176	180	179	175	169
No. of houses with 1 light scheme	22	36	28	33	26	40
No. of sodium lamps	5	6	8	8	6	9
No. of electric pump sets	10	23	42	33	25	36
Total	209	256	278	272	250	269

Source: Computed

road is 1483.95 meters and length of mud road is 1469.52 meters. Sandorpalayam vilage has 4078 meters total length of road and among the length of length tar road length is 1206 meters, length of cement road is 1489.25 meters and length of mud road is 1382.75 meters. Ayekuppam vilage has 4193 meters total lenght of road and among the length of length tar road length is 1455 meters, length of cement road is 1469.00 meters and length of mud road is 1269.33 meters. Passar vilage has 4254.85 meters total lenght of road and among the length of length tar road length is 1500 meters, length of cement road is 1297.85 meters and length of mud road is 1475 meters. L. N. Puram vilage has 3950 meters total lenght of road and among the length of length tar road length is 1096 meters, length of cement road is 1395 meters and length of mud road is 1459 meters.

A study of data in Table 7 indicates the village wise distribution of water infrastructure facilities. It could be noted that Sandorpalayam village takes the first position with respect to possession of water infrastructure facilities. It has 92 water infrastructure facilities. Among them 75 are household tape connections, 2 bridges across the canals, 7 public tape connections and 8 public hand pumps. L.N. Puram village occupies the second position with respect to possession of water infrastructure facilities. It has 84 water infrastructure facilities. Among them 66 are household tape connections, 1 bridges across the canals, 9 public tape connections and 8 public hand pumps. Ayekuppam village occupies the third position with respect to possession of water infrastructure facilities. It has 77 water infrastructure facilities. Among them 60 are household tape connections, 4 bridges across the canals, 8 public tape connections and 5 public hand pumps. Erumpur village takes the fourth position with respect to possession of water infrastructure facilities. It has 74 water infrastructure facilities. Among them 58 are household tape connections, 4 bridges across the canals, 5 public tape connections and 7 public hand pumps. Atarmangalam village occupies the fifth position with respect to possession of water infrastructure facilities. It has 70 water infrastructure facilities. Among them 50 are

Table 6
Village Wise Distribution of Road Facilities

Village wise	*Atarmangalam*	*Erumpur*	*Sandorpalayam*	*Ayekuppam*	*Passar*	*L. N. Puram*
Length of Tar Road	1222	1233	1206	1455	1500	1096
Length of cement road	1366.98	1483.95	1489.25	1469.00	1297.85	1395
Length of mud road	1422.36	1469.52	1382.75	1269.33	1457	1459
Total	4011.34	4186.47	4078	4193.33	4254.85	3950

Source: Computed

Table 7
Village wise Distribution of Water infrastructure facilities

Village wise	*Atarmangalam*	*Erumpur*	*Sandorpalayam*	*Ayekuppam*	*Passar*	*L. N. Puram*
No. of hand pumps	6	7	8	5	9	8
No. of public tape	9	5	7	8	6	9
No. of bridges across the canals	5	4	2	4	3	1
No. of houses with tape connection	50	58	75	60	50	66
Total	70	74	92	77	68	84

Source: Computed

household tape connections, 5 bridges across the canals, 9 public tape connections and 6 public hand pumps. Passar village takes the last position with respect to possession of water infrastructure facilities. It has 69 water infrastructure facilities. Among them 50 are household tape connections, 3 bridges across the canals, 6 public tape connections and 9 public hand pumps.

A study of data in Table 8 indicates the village wise distance of access to infrastructure facilities. Atarmangalam village has an average of 5.56 km distance of access to infrastructure facilities. In this village, distance of access to bus station is 3 km, telegraphic office 5 km, post office 2 km, fair price shop 3 km, college 12 km, secondary school 5 km, nearest town 6 km, primary school 9 km and hospital 5 km. Erumpur village has an average of 5.11 km distance of access to infrastructure facilities. In this village, distance of access to bus station is 3 km, telegraphic office 5 km, post office 3 km, fair price shop 3 km, college 10 km, secondary school 3 km, nearest town 9 km, primary school 5 km and hospital 5 km. Sandorpalayam village has an average of 6.44 km distance of access to infrastructure facilities. In this village, distance of access to bus station is 4 km, telegraphic office 4 km, post office 2 km, fair price shop 3 km, college 23 km, secondary school 4 km, nearest town 5 km, primary school 4 km and hospital 9 km. Ayekuppam village has an average of 8.33 km distance of access to infrastructure facilities. In this village, distance of access to bus station is 5 km, telegraphic office 3 km, post office 5 km, fair price shop 4 km, college 29 km, secondary school 6 km, nearest town 12 km, primary school 2 km and hospital 9 km. Passar village has an average of 6.78 km distance of access to infrastructure facilities. In this village, distance of access to bus station is 4 km, telegraphic office 5 km, post office 3 km, fair price shop 2 km, college 20 km, secondary school 9 km, nearest town 6 km, primary school 6 km and hospital 6 km. L.N. Puram village has an average of 7.22 km distance of access to infrastructure facilities. In this village, distance of access to bus station is 5 km, telegraphic office 2 km, post office 4 km, fair price shop 3 km, college 24

Table 8

Village Wise Distance Access to Infrastructure Facilities

Village wise	*Atarnangalam*	*Erumpur*	*Sandorpalayam*	*Ayekuppam*	*Passar*	*L. N. Puram*
Distance of access to hospitals	5	5	9	9	6	8
Distance of access to primary health centre	9	5	4	2	6	9
Distance of access to nearest town	6	9	5	12	6	5
Distance of access to secondary school	5	3	4	6	9	5
Distance of access to colleges	12	10	23	29	20	24
Distance of access to fair price shop	3	3	3	4	2	3
Distance of access to post office	2	3	2	5	3	4
Distance of access to telegram office	5	5	4	3	5	2
Distance of access to Bus station	3	3	4	5	4	5
Total	5.56	5.11	6.44	8.33	6.78	7.22

Source: Computed

km, secondary school 5 km, nearest town 5 km, primary school 9 km and hospital 8 km.

CONCLUSION

It could be seen clearly from the above discussion that coastal region occupies the first position with respect to high population density, percentage of urban population, percentage of scheduled caste population, average number of households, net irrigated area, net sown area, gross cropped area and per cent of literacy. The flood plain region occupies the first position with respect to high sex ratio, bovine population, draught animal and radio in use. The north plain region occupies the last position with respect to scheduled caste population, workers population, workers in manufacturing industry, cropped area and per capita net sown area. Inter block variation and inter village variation have been observed with respect to rural infrastructure development in Cuddalore district.

REFERENCES

William Nordhaus, and James Tobin (1973), "Is Growth Obsolete", In the Measurement of Economic and Social Performance, Ed., Milton Moss Studies in Income and Wealth, 38, New York, *National Bureau of Economic Research.*

Simon Kuznets (1953), "National Income and Economic Welfare", Reprinted in Simon Kuznets, *Economic Change*, New York: W. W. Norton), 206.

Phiroza B. Medhora (1971), "Poverty in India: A Comment", *Economic and Political Weekly.*

International Labour Organisation (1977), Growth and Basic Needs: A One World Problem, New York, Praeger Publications, 191-102.

Eberts Randall W. (1990), ìPublic Infrastructure and Regional Economic Development,î *Economic Review*, Federal Reserve Bank of Cleveland, 26(1): 15-27.

Henry M., and T. G. Johnson, (1993), *The Contribution of Transportation to Rural Economic Development*, Southern Rural Development Center, Mississippi State University, 171: 35-46.

Forkenbrock David J., Thomas F. Pogue, David J. Finnegan, and Norman S. J. Foster, (1990), Transportation Investment to Promote Economic Development,î in Infrastructure Investment and Economic Development: Rural Strategies for the 1990ís, AGES-9069, U.S. *Dept. Agr., Econ. Res. Serv.*, November.

Kusmin Lorin D., John M. Redman, and David W. Sears (1996), *Factors Associated with Rural Economic Growth: Lessons from the 1980ís*. TB- 1850, U.S. Dept. Agr., Econ. Res. Serv., September.

McHugh Richard J., and James T. Wilkinson, (1988), ìA Random Effects Approach to Substate Growth Models: A Comment on ëThe Determinants of County Growth,íî *Journal of Regional Science*, 28(2): 271-73.

Nijkamp Peter., (1986), ìInfrastructure and Regional Development: A Multidimensional Policy Analysis,î *Empirical Economics*, 11(1): 1-21.

Appalachia, (1982), ìAppalachian Highways Are Catalysts of Change,î 15(2/3): 8-17.

Moon Henry E., Jr., (1987), ìInterstate Highway Interchanges Reshape Rural Communities,î *Rural Development Perspectives*, 4(1): 35-38.

Rephann Terance J., (1993), Highway Investment and Regional Economic Development: Decision Methods and Empirical Foundations,î Urban Studies, 30(2): 437-50.

Rephann Terance J., and Andrew M. Isserman, (1994), New Highways as Economic Development Tools: An Evaluation Using Quasi-Experimental Matching Methods,î *Regional Science and Urban Economics*, 24(6): 723-51.

Siccardi A. J., (1986), ìEconomic Effects of Transit and Highway Construction and Rehabilitation,î *Journal of Transportation Engineering*, 112(1): 63-76.

Cromley Robert G., and Thomas R. Leinbach, (1981) ìThe Pattern and Impact of the Filter Down Process in Nonmetropolitan Kentucky,î *Economic Geography*, 57(3): 208-24.

INDEX

F

Q & R

S